MAKING CONTACT

ING
CONTACT

JILL TARTER
AND THE SEARCH FOR
EXTRATERRESTRIAL
INTELLIGENCE

SARAH SCOLES

PEGASUS BOOKS
NEW YORK LONDON

MAKING CONTACT

Pegasus Books Ltd.
148 W. 37th Street, 13th Floor
New York, NY 10018

Copyright © 2017 Sarah Scoles

First Pegasus Books edition July 2017

Interior design by Maria Fernandez

Library of Congress Cataloging-in-Publication Data is available.

ISBN: 978-1-68177-441-1

10 9 8 7 6 5 4 3 2 1

Printed in the United States of America
Distributed by W. W. Norton & Company

To Earthlings everywhere.

CONTENTS

FOREWORD 1

TERMS 5

PEOPLE 13

ONE HOW'D A NICE GIRL LIKE YOU GET INTO A FIELD LIKE THIS? 23

TWO BABIES, BROWN DWARFS, AND BIG MOVES 45

THREE MAKING THE ALLEN TELESCOPE ARRAY 69

FOUR THE FUTURE OF THE ALIEN-HUNTING TELESCOPE 93

FIVE A QUESTION FOR OUR TIME 105

SIX THE POLITICS OF SCIENCE AND NEW PROJECTS 133

SEVEN THE QUEST FOR CONTACT 161

EIGHT THE LAST CHAPTER 183

NINE EXTREMOPHILES AND EXOPLANETS 207

TEN SHOUTING INTO THE VOID 227

A TIMELINE OF SETI 259

FURTHER READING 263

ACKNOWLEDGMENTS 267

INDEX 269

FOREWORD

During a conference celebrating the fiftieth anniversary of the search for extraterrestrial intelligence (SETI), Jill Tarter left her purse on a bus—a bus I happened to be driving. At the time, I was fresh out of graduate school, working at my first post-education job as a public outreach officer at the Green Bank Telescope. On that particular night, my job involved driving a large vehicle full of SETI dignitaries to the telescope so they could eat dessert underneath its dish, then driving them back, and never crashing.

I parked the bus and ran after Tarter, who was then the director of the SETI Institute and whose fictional alter ego, a character from the book and movie *Contact*, had been my childhood inspiration for entering the radio astronomy world.

"I think you left your purse," I said when I reached her. I extended my arm, the bag hanging awkwardly from it like I was a person who didn't carry bags.

She said, "Thank you," smiled, and continued toward the observatory's lounge.

In that same lounge, astronomers had held the first SETI conference a half-century earlier. They gathered under the rainbow light of a 1960s chandelier and debated this: Are we alone in the universe? And if we're not, how much company do we have? They poured the foundation for the SETI work that would come, stacking itself atop their conclusions, asking over and over those same initial questions and investigating them with new scientific knowledge and technology.

As I watched Tarter walk away, I thought about how I could have recited her resume to her, having watched her career ever since I'd known what a career was. That milestone of mental development coincided with my watching of the movie *Contact*, based on a 1985 book by Carl Sagan about the discovery of alien life, when I was twelve. I was enraptured in a way I hadn't ever been before, even though I'd been reading dubious but thrilling books about wormholes and black holes and time travel for years. Here, in SETI, was this whole scientific world that I didn't know existed, and here was this person leading its charge—a fierce, determined, stubborn, smart woman who asked big questions about the universe and didn't hear "No" as "No" but as "Keep trying." I wanted to be like her, to investigate the kinds of mysteries she investigated, to know the things she knew.

I watched the film obsessively, mouthing, "If it's just us, that seems like an awful waste of space," along with Jodie Foster, who played the extraterrestrial finder Ellie Arroway. I read and reread Sagan's book, glossing over my inability to understand ruby masers, obsessing instead over my cosmic insignificance and the broad brushstrokes of the big cosmic questions.

At my first college astronomy internship, with the National Radio Astronomy Observatory, I lived in a house with five other interns.

One night, we watched *Contact* together. Here's why: it was the thing that had brought us here. And that's something I've heard from many astronomers around my age, especially the women: *Contact* was their earliest inspiration, the first depiction they saw of what astronomy was, and the first time they saw a scientific character and thought, "I see myself in that person."

One of the other interns said, "Did you know she's partly based on a real scientist?"

That real person, the intern continued, was a woman named Jill Tarter.

Fictional *Contact* follows Arroway as she searches for a radio signal from an intelligent extraterrestrial civilization, battling bureaucracy, politicians, economic woes, statistical unlikelihood, institutionalized sexism, and her own emotional demons. As a nonfictional woman scientist and a SETI scientist, Tarter faced the same challenges. But this is where the two women's stories depart: Arroway finds a signal. E.T. calls. E.T. sends instructions for building a spaceship. Humanity builds the spaceship (not without trials), and (not without trials) Arroway becomes the sole passenger. To my knowledge—and I've interviewed her a lot—Tarter has never been on a spaceship. But if our real world does make contact with extraterrestrial intelligence— a discovery that would be one of the biggest and most impactful humans have ever made—we will owe much of the thanks to the real Tarter, who got in on the early days of the field, kept it alive during its lean years, and pushed its science and technology forward, even as politicians, the public, and even other scientists said the endeavor was not worthwhile.

After I returned Tarter's purse that night in Green Bank at the SETI conference, I biked home, to astronomer Frank Drake's former bachelor pad—a farmhouse that, set up against the woods and complete with a chicken-slaughtering room in the basement, many found remote and creepy. But I loved it, and I loved knowing Drake had lived in this house when, in 1960, he made the first modern search for extraterrestrial intelligence, called Project Ozma.

He didn't find what he was looking for. That night, I looked up at the bright swath of Milky Way stars and wondered when, or if, anyone ever would.

Three years later, when I was an editor at *Astronomy* magazine, I wrote Tarter an email, informing her that we'd met for five seconds when she lost her purse and then asking if I could write a biography of her. After a bit of back and forth, including the question "Why should I believe you can write a whole book?" she agreed.

Today, I've had the privilege of doing countless hours of interviews with Tarter, of asking her questions I haven't even asked my own family members, of getting to know her in a way that has made her a regular person to me and not that soft-focus, pedestaled hero she was to me when I was younger. We've torn apart file cabinets and photo albums; I've pestered her peers about her personality; I've slogged through transcripts of congressional meetings and 300-page conference reports—all with the idea of understanding how Tarter came into herself, how she came to SETI, how SETI has evolved since becoming a science, and how its future may play out.

Tarter has spent more than 40 years trying to answer the question "Are we alone?" And never during those decades did she know how close she may have been. The wondering keeps her up at night. But her scientific questions are, in some ways, the same as the existential questions that keep us all up at night. We have all spent dark hours wondering about our place in it all, our aloneness both terrestrial and cosmic. Tarter's life and her life's work are not just a quest to understand life in the universe: they are a quest to understand *our* lives in the universe.

So, spoiler alert: this book will not tell you whether They Are Out There. But hopefully it gives some insight on earthlings, and what we're doing down here.

TERMS

100-Year Starship: A research program sponsored by the Defense Advanced Research Projects Agency launched in 2011. Its mission is "to make the capability of human travel beyond our solar system a reality within the next 100 years."

Active SETI: The act of broadcasting messages toward space with the idea that they may be received by extraterrestrial civilizations.

Allen Telescope Array: A SETI-centric telescope made of 42 radio antennas. The SETI Institute and the University of California, Berkeley, completed this scaled-down version of the array in 2007, primarily with funds from the Paul G. Allen Foundation, hoping to expand it eventually to 350 antennas. After Berkeley backed out, leading to a hibernation in 2011, the organization SRI International took over management.

Ames Research Center: The NASA center located in Silicon Valley and the place within the agency where SETI got its start. The center is now a leader in astrobiology and the search for exoplanets.

Arecibo Observatory: The home of the 305-meter-wide telescope on which Tarter and colleagues have done SETI observations for NASA's High Resolution Microwave Survey and the private Project Phoenix. Some of the data that feeds the University of California, Berkeley, SETI@Home program also comes from this telescope.

Band: A range of frequencies or wavelengths.

Bandwidth: The range of frequencies or wavelengths to which an instrument is sensitive.

Blueshift: The apparent squishing of electromagnetic waves into a shorter wavelength that occurs when an object is moving toward the observer.

Breakthrough Initiatives: The Breakthrough Listen project to do SETI, the Breakthrough Message program to send a missive to space, and the Breakthrough Starshot idea to send a suite of small probes to the nearest star. The initiatives are all funded by Russian billionaire Yuri Milner.

Contact: A 1985 science fiction book by astronomer and public figure Carl Sagan that portrayed a successful SETI project. The fictional astronomer who found the fictional aliens was, in part, inspired by Jill Tarter. In 1997, director Robert Zemeckis adapted the novel into a movie of the same name, starring Jodie Foster.

Deep Space Network: A NASA-operated network of radio telescopes that primarily downloads data from and uploads commands to spacecraft beyond Earth's orbit. SETI scientists have also used them to test equipment.

Doppler effect: The redshifting or blueshifting of electromagnetic waves (be they in the radio, infrared, optical, ultraviolet, or gamma range) that happens when the object emitting the light is moving away (redshift) or toward (blueshift) the observer (usually a telescope).

Drake equation: An equation that SETI pioneer Frank Drake created as a kind of agenda for the secret Order of the Dolphin meeting in Green Bank in 1961. The equation considers the factors that would lead to intelligent, technological, communicative life in the Milky Way and multiplies them together to estimate the number of communicating civilizations that likely exist in our galaxy at any given time.

Electromagnetic Radiation: The spectrum of light, made of photons, from radio waves to gamma rays.

Extremophile: An organism on Earth that exists in extreme conditions that most life would find unpalatable.

Exoplanet: A planet outside our solar system.

Feed: The part of a radio telescope that funnels the radio waves from space toward the instrument that actually detects them.

First SETI Protocol: More formally called the "Declaration of Principles Concerning Activities following the Detection of Extraterrestrial Intelligence," this protocol describes the steps scientists and political leaders should take to verify and react to a potential message from extraterrestrials.

Frequency: The number of light waves per second that passes by a given spot, measured in units like Hertz and megaHertz.

FUDD: The Follow-Up Detection Device, used by the SETI Institute. The device—identical copies of which lived at the main telescope and a remotely operated backup telescope—analyzed candidate signals in real-time and could provide evidence that a signal came from space and was not interference from earthly devices.

Green Bank: An observatory site in remote West Virginia. Frank Drake's first SETI project took place here in 1960, as did the Order of the Dolphin SETI meeting the next year. Tarter did early SETI work there, and she and her colleagues used the 140-foot telescope for Project Phoenix. Today, the Breakthrough Listen project runs partly on the 100-meter-wide Green Bank Telescope.

Habitable Zone: The region around a star in which a planet can harbor liquid water.

Hat Creek: The town and observatory site of the Allen Telescope Array.

HRMS: The High Resolution Microwave Survey, NASA's first full-scale SETI project, which began observing in 1992 with the Arecibo Observatory. In 1993, Congress canceled the project.

Jodrell Bank: An observatory in Lower Withington, England, at which Tarter and colleagues performed initial SETI experiments, and where Bernard Lovell secretly used the SETI team's equipment to search for signals from Russian spacecraft.

JPL: NASA and Caltech's Jet Propulsion Laboratory, located in Pasadena, California. During the High Resolution Microwave

Survey, which involved deeply targeting specific stars and doing shallower but wider surveys, this site led the "survey" part of the project.

Kepler Space Telescope: A planet-hunting telescope, launched in 2009, that has found thousands of planets outside our solar system.

Light-Year: The distance light travels in a year, or 5.9 trillion miles.

Maser: A word that stands for "microwave amplification by stimulation emission of radiation." It is essentially a laser—naturally occurring or made by a person—that emits photons at a very, very narrow set of wavelengths (like the single, pure color of a laser) but at the lower energies of radio waves, microwaves, and infrared waves. Tarter and colleagues investigated masers in the universe because they believed they represented the narrowest-band signals nature could produce, so anything narrower had to come from a synthetic source.

METI: Messaging Extraterrestrial Intelligence, another term for "active SETI."

Mobile Research Facility: The mobile tractor-trailer out of which the SETI Institute ran the High Resolution Microwave Survey and Project Phoenix operations at each site.

Moore's Law: The observational idea, put forth by Gordon Moore in 1965, that the number of transistors on each square inch of an integrated circuit doubles every year. A decade later, he revised the idea to a doubling every two years. In both forms, it predicts the exponentially growing power of computers. That growth, though, cannot continue at the same pace forever, although it has held relatively true so far.

MOP: Microwave Observing Project, the planning-stage precursor to NASA's High Resolution Microwave Survey.

Mopra: The second telescope Tarter's SETI team used remotely during observations with Australia's Parkes telescope.

Narrowband: A signal that exists over a small range of frequencies, like a radio station broadcast.

NASA: The National Aeronautics and Space Administration.

NRAO: The National Radio Astronomy Observatory.

Order of the Dolphin: The code name for the group of 10 scientists that gathered in 1961 for a SETI meeting in Green Bank, West Virginia, after Frank Drake's first experiment.

Parkes Radio Telescope: The 64-meter-wide radio telescope in Australia where Project Phoenix began after the scientists were allowed to purchase time on the telescope. Today, Breakthrough Listen uses the telescope for SETI work.

Project Cyclops: A NASA-sponsored study into the kind of telescope that could realistically search for an alien civilization, released in 1971. The committee proposed an array of many small radio antennas that work together, as the Allen Telescope Array's antennas do today. NASA never built this proposed telescope, but the report from the project first inspired Tarter to join SETI as an enterprise.

Project Ozma: Frank Drake's first SETI project, in which he used the 85-foot Tatel telescope in Green Bank to search for radio signals from the stars Tau Ceti and Epsilon Eridani in 1960.

Project Phoenix: The private SETI project that "rose from the ashes" of NASA's canceled High Resolution Microwave Survey. It ran from 1995 to 2004, using the telescopes at Parkes, Green Bank, and Arecibo.

Receiver: The part of a radio telescope that collects and detects radio waves, analogous to a camera at the focus of an optical telescope.

Redshift: The apparent stretching of electromagnetic waves into a longer wavelength that occurs when an object is moving away from the observer.

RFI: Radio frequency interference, or human-produced radio waves—from things like Wi-Fi routers, cell phones, digital cameras, satellites, and spark plugs—that appear in and contaminate radio telescope data.

Second SETI Protocol: A proposed document detailing how to decide whether to purposefully broadcast a message to space, with the idea that it could reach an alien civilization, and the best practices for such a broadcast.

SERENDIP: An ongoing set of SETI projects named the "Search for Extraterrestrial Radio Emissions from Nearby Developed Intelligent Populations." Tarter worked on the initial version, and the Berkeley-run project is now on its sixth incarnation.

SETI Institute: The institute that Tarter co-founded to study all aspects of the existence, formation, and evolution of life in the universe. It came into being in 1984, and it continues today in Mountain View, California, employing scientists who research topics ranging from exoplanets to extremophiles to Martian geography. The SETI Institute also operates the Allen Telescope Array.

SETI Live: A short-lived citizen science program that allowed users to categorize signals from real-time SETI data delivered from the Allen Telescope Array. The funding for the project came from Tarter's 2009 TED Prize.

SETI Quest: The TED Prize–sponsored portal that allowed members of the public to participate in the SETI Institute's work, in the form of software and algorithm development and citizen science.

Wavelength: The distance between two peaks of an electromagnetic wave. Radio waves (including microwaves) range from about 1 millimeter to 100 kilometers.

PEOPLE

Peter Backus: Backus was a radio astronomer and programmer for NASA's SETI program from 1985 to 1988 and its co-principal investigator from 1988 to 1994. After the program was canceled, he worked in management positions for Project Phoenix, as a software engineer for the SETI Institute's search system and as a manager of the SETI Institute's SETI observing programs. Today, he is a member of the NASA Astrobiology Institute.

Natalie Batalha: Batalha is an astrophysicist at Ames Research Center and serves as mission scientist for the planet-hunting Kepler Space Telescope, which she has worked on since it existed in only proposal form. She has been awarded a NASA Public

Service Medal and is a leader on the agency's Nexus for Exoplanet System Science program to understand which exoplanets may be habitable.

John Billingham: Known as the "father of SETI at NASA," Billingham co-directed Project Cyclops and, afterward, began NASA's SETI program. According to his 2013 obituary at the SETI Institute's website, he "transformed SETI from an occasional experiment into a systematic program."

David Black: Black served as CEO of the SETI Institute from 2014 to 2015. His previous positions include president and CEO of the Universities Space Research Association, chief scientist for the International Space Station, and deputy chief for the Space Sciences Division at Ames Research Center.

William Borucki: After designing heat shields for the *Apollo* spacecraft, Borucki masterminded the Kepler Space Telescope, championing its concept for decades before it became reality. He has received the National Academy of Sciences' Henry Draper Medal and the Shaw Prize for his efforts in finding exoplanets.

Stuart Bowyer: Bowyer, an astronomer at the University of California, Berkeley, is credited with founding the field of extreme ultraviolet astronomy. Tarter met Bowyer at Berkeley, where he showed her the *Cyclops Report* that inspired her to begin her SETI work. He headed the SERENDIP program on which she got started in the field.

David Brin: Brin is a scientist, futurist, and science fiction author who has won the Hugo, Nebula, Locus, and Campbell awards for his writing. In the field of active SETI, he is a strong voice in opposition of broadcasting messages without significant forethought and oversight.

Richard Bryan: Bryan served as a Democratic senator from Nevada between 1989 and 2001. In 1994, he introduced an amendment that killed the nascent NASA SETI program, the High Resolution Microwave Survey.

Giuseppe Cocconi: Cocconi, a physicist, authored the seminal *Nature* paper "Searching for Interstellar Communications" in 1959, suggesting scientists could look for communications from extraterrestrials by doing radio searches around the wavelength at which hydrogen atoms naturally emit radio waves.

Betty and Dick Cornell: Jill Tarter's mother and father, who raised her up into the scientist and engineer she is today.

Kent Cullers: Cullers worked as a team leader and subsystem manager for the targeted search portion of NASA's first SETI project, on which he was also key to the digital data processing effort. At the SETI Institute, he was a senior scientist and project manager for Project Phoenix. The blind astronomer, Kent, in the movie *Contact* is based on this real Kent.

David DeBoer: DeBoer is a research astronomer at the Radio Astronomy Lab at the University of California, Berkeley, where he works on, among other things, the Breakthrough Listen initiative. In the past, he has served in leadership positions for the Allen Telescope Array and helped set up the Woodbury telescope for Project Phoenix.

Bill Diamond: Diamond began a position as president and CEO of the SETI Institute in 2015, after spending decades in the world of industry, dealing with photonics, optical communications, X-ray technologies, and semiconductors.

Frank Drake: Drake conducted the first modern search for extraterrestrial intelligence, called Project Ozma, in 1960. He authored the

Drake equation and was a key figure at the Order of the Dolphin meeting, the first SETI conference. He also sent the first radio broadcast containing a message meant for extraterrestrials—the Arecibo message. Drake was the president of the SETI Institute from 1984 to 2000 and later was the director of its Carl Sagan Center for the Study of Life in the Universe.

John Dreher: Dreher became part of NASA's High Resolution Microwave Survey, its first SETI project, in 1989 and continued his SETI work on Project Phoenix when the federally funded search ended. Dreher also acted as project scientist and program manager for the Allen Telescope Array during its research and development period and continued as project scientist during its construction, while also working on its detectors.

Ann Druyan: Druyan co-wrote the original *Cosmos* series, starring Carl Sagan, and created, produced, and co-wrote the 2014 redux, *Cosmos: A Spacetime Odyssey*, staring Neil deGrasse Tyson. She was also the creative director for the Voyager Interstellar Message project, which placed golden records—etched with a multimedia portrait of humanity—aboard the twin *Voyager* spacecraft. She and Sagan married in 1981.

Ron Ekers: Ekers has been the president of the International Astronomical Union, the director of the Very Large Array, and the foundation director of the Australia National Telescope Facility. It was in this latter role that he rented the Parkes telescope to the Project Phoenix team.

Debra Fischer: Fischer is an exoplanet astronomer at Yale University. She is currently developing a next-generation instrument called ExPRESS to find planets like Earth. She has discovered hundreds of planets—as well as the first exoplanetary system that contains multiple worlds, a discovery she made in 1999.

Jake Garn: A Republican senator from Utah who championed SETI in government meetings during the early 1990s.

Daniel Goldin: Goldin served as NASA administrator from 1992 to 2001. It was at the beginning of his tenure that Congress canceled the agency's SETI program. He is currently the president and CEO of hardware company KnuEdge.

Sam Gulkis: Gulkis co-led the survey portion of the High Resolution Microwave Survey from the Jet Propulsion Laboratory in Pasadena, California.

Gerry Harp: Harp is currently co-director of the SETI Institute's Center for SETI Research. He has worked at the institute since 2008, coming to the organization with a background in quantum mechanics.

Mae Jemison: In 1992, Jemison became the first African American woman to go to space. An astronaut, medical doctor, and engineer, Jemison currently heads the DARPA-sponsored 100-Year Starship program, which aims to pave the way for future interstellar travel.

Yuri Milner: A Russian billionaire, Milner founded the Breakthrough Prizes—awards in physics, life sciences, and math—as well as Breakthrough Initiatives, which include the Breakthrough Listen program to search for intelligent extraterrestrials, to learn more about life in the universe.

Philip Morrison: A former Manhattan Project physicist, Morrison co-authored the 1959 *Nature* paper "Searching for Interstellar Communications," suggesting scientists could look for communications from extraterrestrials by doing radio searches around the wavelength at which hydrogen atoms naturally emit radio waves.

Chris Neller: A research assistant and admin at the SETI Institute, Neller was the reason Tarter kept track of her travel and paperwork.

Bernard Oliver: Oliver, nicknamed "Barney," led research and development at Hewlett-Packard for approximately 40 years. He traveled to Green Bank, West Virginia, to see Frank Drake's early SETI efforts and attended the Order of the Dolphin conference shortly thereafter. He later co-directed Project Cyclops and became senior manager for NASA's SETI program. When Congress canceled this federal project, he helped the SETI Institute secure private funds to implement the private Project Phoenix, for which he was a volunteer senior scientist.

Tom Pierson: Pierson co-founded the SETI Institute and was the one who filed its official paperwork in 1984. He served as its CEO for decades afterward, growing it from its initial small staff, focused on SETI, to a multidisciplinary organization of more than 100 researchers.

William Proxmire: Proxmire served as a Democratic senator from Wisconsin from 1957 to 1989. He was an outspoken opponent of SETI and awarded NASA's efforts to find life in the universe with his Golden Fleece Award—issued for what Proxmire deemed wasteful government spending—in 1978.

Jon Richards: Richards runs the Allen Telescope Array's observing program and keeps the public updated in real time about what's going on at the telescope at SETIquest.info.

Carl Sagan: What *didn't* Sagan do? He was the host of the television series *Cosmos*; the author of the book *Contact*, which features a protagonist partly modeled on Tarter; and one of the masterminds of the Voyager golden records. He wrote many popular books and hundreds of scientific papers and championed SETI, and science in general, in the public arena. Decades after his death, people still name him when asked to think of a famous astronomer.

Seth Shostak: Shostak is currently co-director of the SETI Institute's Center for SETI Research. He hosts the podcast *Big Picture Science* and authored the book *Confessions of an Alien Hunter*.

Bruce Tarter: Tarter's first husband, Bruce was the director of the Lawrence Livermore National Laboratory from 1994 to 2002.

Shana Tarter: Tarter's daughter and the assistant director of the Wilderness Medicine Institute at the National Outdoor Leadership School.

Charles Townes: Townes, the inventor of the maser and the laser, won the Nobel Prize in physics in 1964. He was the first to come up with the idea of optical SETI.

Margaret Turnbull: Turnbull is an astrobiologist and exoplanet astronomer who, with Tarter, developed the HabCat, a catalog of star systems that might be friendly to life.

Douglas Vakoch: Currently the president of METI International, a nonprofit organization focused on active SETI, Vakoch was previously the director of interstellar message composition at the SETI Institute. He is the head of the International Academy of Astronautics Study Group on Active SETI: Scientific, Technical, Societal, and Legal Dimensions.

Jack Welch: Welch, Tarter's second and current husband, is considered a founder of molecular radio astronomy. He has directed the Radio Astronomy Lab at the University of California, Berkeley; has held the university's Watson and Marilyn Alberts Chair in the Search for Extraterrestrial Intelligence; and was key to the development of the electronics inside the Allen Telescope Array antennas. The National Academy of Sciences, of which he is a fellow, says he "started the field of millimeter-wave interferometry."

Dan Werthimer: The chief scientist for SETI at the University of California, Berkeley, Werthimer leads its SETI@Home project and the ongoing SERENDIP investigations. He is also now a key leader of the Breakthrough Listen program. When Berkeley and the SETI Institute worked together on the Allen Telescope Array, Werthimer was also part of that program.

Pete Worden: Worden is the former director of NASA's Ames Research Center and the current chairman of Yuri Milner's Breakthrough Prize Foundation.

The Universe is a pretty big place. If it's just us, seems like an awful waste of space.

— Carl Sagan, *Contact*

CHAPTER ONE

HOW'D A NICE GIRL LIKE YOU GET INTO A FIELD LIKE THIS?

O n July 31, 2014, Auditorium 220 at NASA's Ames Research Center fills with employees—a mix of 10-year-old button-down shirts, pleated Dockers, and the designer jeans of hip postdoctoral researchers. Pete Worden, the center's director, steps in front of the crowd to introduce the afternoon's speaker. Astronomer Jill Tarter stands to the side of the stadium seats, ready to deliver a talk called "Searching for ET: An Investment in Our Long Future." She wears rimless glasses and the big bronze turtle earrings that go everywhere with her, her hands clasped behind her back. As Worden lists Tarter's accomplishments—all related to her 40-year involvement in the search for extraterrestrial intelligence (SETI)—she shakes her head and smiles, laughing when he says, "Most of us are still trying to find intelligent life here on Earth." When Worden finishes and Tarter takes the stage, the audience lifts their smartphones to snap pictures of her in front of an introductory PowerPoint slide before the lights dim.

"Your story—my story, our story—began billions of years ago," she says. "But that probably doesn't come first to your mind when you wake up in the morning." A pause fills the air-conditioned room. "We need to change that."

The screen flashes to an illustration of the Big Bang, 13.8 billion years ago, and then Tarter clicks to an image of a pinwheeling galaxy like our Milky Way, born 10 billion years ago.

"We are intimately connected with these faraway times and faraway places," she continues, "because it takes a cosmos to make a human."

People nod in assent, feeling like they do when a book's narrator articulates a fundamental truth that they themselves have never molded into words.

It has taken humans millennia to even begin to figure out how that cosmos made us, and we're still not sure. We know a bit about how the universe began, how galaxies and stars got their starts, why planets exist, and what ours was like when the first microorganism claimed its spot in that strange ecosystem. But we don't understand how earthly chemistry spat out that first living thing, or whether similar sequences have swept across other planets.

"We want to know," Tarter says, "where do we come from? Where are we going? What is? *Why* is? And of course, we're really interested in whether or not there's anyone else out there."

She shakes the projection system's clicker, which is responding on a delay, like it's communicating with a device light-years distant. "This thing is a bit jet-lagged," she says, and the audience laughs.

Given that we can't even get PowerPoint to work properly, it's clear that our civilization is very young, technologically speaking. And we don't know how long it will survive. We could cause our own demise (climate change, nuclear war, antibiotic-resistant epidemic), or death could come from above (large asteroid on a collision course, nearby supernova, dying sun). But for how young humanity—and all Earthling life is—Earth resides in an ancient galaxy. And odds are, if we find anyone else inside, they will be much older than we are (on the cosmic timescale, if they were any younger, they would still be excited about discovering fire).

Tarter's face shifts as she becomes frank and colloquial. "Look," she says, "I don't expect extraterrestrial salvation. I don't expect them to tell us how to solve our problems. But the very existence of such a civilization should motivate us to figure out a way to solve our own problems."

Maybe SETI will find such alien civilizations. Their distant and unimaginable lives will convince us that we, too, can avoid blistering our planet or blowing each other up. Maybe humans, or whatever we evolve into, could live long enough to build interstellar transit systems or some other sci-fi-seeming technology.

But maybe SETI will fail. Maybe E.T. won't phone home. Maybe there is no E.T.

Tarter, who has spent nearly her whole adult life hoping the cosmic phone will ring, would of course be disappointed. But Tarter claims, in this speech, that the search will be worthwhile (actually, she says "really, really, really worthwhile," emphasizing with an up-and-down gesture of her hand and closing her eyes), even if we don't find any aliens. SETI holds up a mirror, showing us how we look from a cosmic perspective—a perspective that began 13.8 billion years ago and encompasses 2 trillion galaxies beyond our own.

"And in that mirror," she says, "we are all the same. It has the effect of trivializing the differences among Earthlings, differences that we're willing to spill blood over. We have to get over that. I think SETI is a great way to do it."

❖

The first search for extraterrestrial intelligence happened in 1960, in Green Bank, West Virginia, when Tarter was still in high school. Astronomer Frank Drake, a young researcher with the National Radio Astronomy Observatory, had a bold plan and approval from the facility's director. Drake had calculated that the observatory's 85-foot radio telescope, which looks like a gigantic version of the old-timey satellite dishes still rusting in rural yards, could detect extraterrestrial radio broadcasts as weak as the ones humans then transmitted. No one had ever sought out extraterrestrials before, and it was possible they were abundant and talkative, just waiting for us to tune in. For all Drake—or anyone—knew, every star was home to a thriving civilization, blasting the *Encyclopedia of Everything You Want to Know about the Universe* into space and hoping societies like ours would pick up the transmission. It was a heady time, filled with big questions and potentially bigger answers. Are we alone? How did we get here? Where are we headed?

Drake selected two nearby sun-like stars, pointed the telescope at them, and scanned through frequencies much like you do when searching for an FM radio station. He sat alone in the buzzing control room beneath the telescope dish. As the night sky spun on, he wondered whether when he looked up at those two stars, called Tau Ceti and Epsilon Eridani, someone—maybe with two legs and a radar transmitter, or four legs and a broadcaster he couldn't yet dream of—was looking out from their own planet at a tiny dot in their sky: our sun.

Over the course of four months, Drake sat in that control room for 115 hours, listening and hoping for signs of that someone.

He heard nothing.

❖

Five hundred miles away, in suburban New York, a girl named Jill Cornell scrawled diligently away on her physics homework. Cornell, who later became Jill Tarter, first considered the possibility of alien life on a family vacation to visit her aunt and uncle on Florida's Manasota Key. Her relatives had once been Manhattan bankers, but they got tired of the rat race and the nepotism rules that meant their marriage was a secret. So they chased the other American dream—to be free from the churn of chasing the American dream—and chucked their careers. They moved south to be beachcombers and construction workers. They built a succession of coastal houses, living in each before selling it and moving on. While the key is now filled with Airbnb vacationers clutching margaritas, Manasota once was nearly uninhabited, even lacking power lines during young Tarter's earliest stays there. Without light pollution, the island sky seemed as dark as space actually is, Sagittarius for once looking like a convincing archer.

When Tarter was 10, her father took her for a seaside walk to teach her the constellations. He pointed up at each set of stars, explaining how they seemed to connect into a coherent picture. But they were actually light-years apart, he told her, only by happenstance lining up in our sights. Standing on the deserted beach with her father, toes knocking against the seashells she collected and categorized during the day, she considered the idea that those stars might be someone else's suns. It made sense to her. And on some planet circling one of those stars, some other creature was probably walking along some other coast with her (his, its) parents. They could peer up at their sky and see our sun, which would be a part of constellations unrecognizable to us.

This is all very romantic—the mythology of a person, a convenient story in which the arc from past to present is clear. But such a hindsight-hued vision of history simplifies the way life actually is. Back then, Tarter was just a kid daydreaming about aliens. There is, really, nothing special about that. I did that; you probably did that. Sure, Tarter was smart, interested in exploring the universe from an early age (envious, for instance, that Flash Gordon got to drive the spaceship while his female accomplice, Dale Arden, waited around in a short skirt). But no one could have crowned Jill Cornell the future queen of SETI back then. There was no way to know she would eventually emerge from the crowd of other kids who had the same ideas.

But because she *did* emerge from the crowd of other kids who had the same ideas, the beach scene seems different—filled with foreshadowing and prelapsarian mythos. Before there were committee meetings and congressional hearings, before telescope groundbreakings and technology mishaps, before fundraising pitches and receiver upgrades, there was only this walk. There was only this childhood certainty that the scene was duplicated, triplicated, even quadruplicated elsewhere.

❖

Tarter grew up on the water—on Florida vacations as well as at home, in a small lakeshore apartment in Eastchester, New York. As an only child, she spent most of her conversational energy talking to adults, and her parents' full focus fell on her. She lit up under her father's beam of attention. He fostered their similarities—a love of deciphering the innards of objects, a desire to pull fish out of water, and a sense that the universe could be grabbed and dragged down to Earth for examination.

After he fought in World War I, Jill's father, Dick, attended Swarthmore College. When the school offered him a scholarship to stay and study graduate-level astronomy, he declined, becoming a professional football player instead. Tarter has newspaper clippings, found in a family trunk that once belonged to Dick's mother, from his athletic career. The yellowed pictures, whose ink has bled into three generations of fingertips, show him in 1920s sports gear. The kneepads are flimsier than inline skaters', and the whole ensemble looks like a traumatic injury waiting to happen. But Dick himself looks tall, strong, charming—like he didn't need kneepads anyway. Like if he told you something, you would believe it, or would at least want to.

Dick wanted his daughter to be self-reliant, and so he set milestones for her to achieve before she could do whatever it was she wanted to do. He let her go canoeing with him only if she portaged the boat, and she was allowed to attach herself to his hunting and fishing trips only if she braided her own hair first (a task that Jill's mother, Betty, had performed for years, but one for which Dick had no time or skill). By the time she was eight, she had mastered the art of hair engineering and grown boat-hefting muscles, so she spent much of her free time with her father and his friends in sportsmen's camps. She caught snakes, fish, and the equivocal attention of her mother, Betty.

"What would *they* think?" her mother asked her.

Who are they? Tarter thought. *And what do I care?*

Betty informed Dick that their daughter was not developing the way a girl should. She wasn't graceful. She had a habit of draping reptiles across her shoulders. But she was only eight, so a social 180 was still possible. The next day, Dick hoisted Tarter and set her on the washing machine, which was her mother's most prized possession and which occupied a large portion of the real estate in the kitchen. Whenever Dick sat her down there, Tarter knew they were going to have a *conversation*. This perch put her at eye level with her father, and she looked at his irises, waiting.

"Your mother thinks," he said, in the timeless way of fathers, "that maybe since you're getting older, you should be spending more time with her, rather than me."

Tarter waited.

"Learning how to do girl things," he continued. Whether or not he believed his own words, he insisted: this is what it means to grow up.

"Why do you have to do one or the other?" she asked. "Girl things or boy things?"

Dick had answers, sociological ones that Tarter didn't want to hear. "These are the ways of the world," he said.

Luckily, having spent so much time with her father, she knew how to manipulate him. She combined tears and logic, batting back rebuttals. Rhetoric plus emotional reaction yielded the desired product.

"Well, as long as you're willing to work hard enough," he relented, "I don't see why you couldn't do anything you want."

His cave-in opened up an escape route, a way into the world she wanted to inhabit. Tarter calls what she said next the Washing Machine Declaration. What was the most male thing she could be, besides an actual boy? She knew from hanging around the sportsmen's camps that lots of those guys were engineers. And although she had only vague notions of what an engineer did all day, she hardly paused before proclaiming, "I want to be an engineer."

"And that was that," Tarter says now. "From that moment on, that was just what I was going to do."

She hopped off the washing machine.

❖

If she was going to be an engineer, her father was going to help her act like one. One day, he handed her a plastic transistor radio.

"Take it apart," he said. "And then put it back together."

Cracking open its shell, she found a whole world of indecipherable circuits, ribbons, and resistors, all communicating with each other. Finding and then dismantling that world was easy, but reconstructing it was not. There were pieces left over, bits of metal that seemed not to fit anywhere. She called for her father in a childish voice—the kind that adds an extra syllable to "Dad"—wanting him to make the inscrutable parts make sense again.

"You need to figure it out yourself," he said.

She did, in short order.

"I was fearless when he was around," she says. "He was the center of my universe."

But that center wouldn't hold.

"Yes, he was ill," she says, of the cancer that killed him just a year or two after he gave her the radio. "Yes, he was in and out of hospitals with cancer, and I had no idea how serious it was. But of course he was always going to be there."

Given her grasp on evidence in the rest of her life, it's clear that her emotional universe was subject to different laws from the physical one. It's a characteristic that holds to this day, an almost "law of attraction" philosophy: If she believes something won't, or will, happen, it won't (or will).

In spite of Tarter's denial, Dick died when Tarter was 12, just four years after the Washing Machine Declaration. His death shook her like a high-Richter earthquake—totally devastating, seemingly illogical, and completely unexpected. The ground was supposed to stay firm underfoot. But now that her father was gone, she knew familiar terrain could shift without warning, or rip itself apart. Her universe, the universe, became a centerless thing, vast and hard to comprehend.

Dick's death rooted one particular idea in Tarter's brain, an idea that would go on to have as much effect on her life as any academic degree or astronomical insight. The day he died, she remembered a question she had wanted to ask him. She doesn't remember what it was—it wasn't anything particularly important. "It was just this realization of a big black void," she says. "I couldn't ask him *anything* anymore."

I should have asked him yesterday when I had the chance, she thought, the lesson stitched painfully into her brain.

When she was older and knew more Latin, she recognized the lesson as carpe diem. If you have an opportunity, take it. Now. Or else it may sublimate from something concrete into something absent.

Her father's death also transformed the Washing Machine Declaration into a compass that oriented her life for years. "I told my dad I was going to become an engineer," she says. "And I wanted to make my dad proud." Imitating her determination those 60 years ago, she continues, "'Goddammit, there's no way I'm going to wimp out.'"

She probably used such profanity then, too, despite what *they* may have thought.

❖

Tarter began researching what engineers did all day, and what sort of an education a person needed to do whatever that was: physics and high-level math in high school, she found, courses that fell into the "it's just not done" category for women, even if the classroom doors lacked NO GIRLS ALLOWED signs.

"The guidance counselors of the era were just so awful," Tarter says of high school in the late fifties and early sixties. "'Why do you want to take calculus? You're just going to get married and have babies.'" (She, in fact, did get married and have a baby. But it turns out that physics was—wonder of wonders—still useful and that one endeavor did not mutually exclude the other.)

Her physics teacher, whom the students called Doc, looked like her father. That similarity lit up her brain in places left dark after her father died. This interpersonal déjà vu led her to trust him with an extracurricular problem.

The fashion accessory du jour was a chicken wishbone, strung through a chain and worn as a charm. Suburban students walked around looking as if they might break into a pagan ritual at any time. Tarter wanted to silver-plate hers. A metal-wrapped wishbone never breaks. She would never have to battle for the long end, never herself get the short end. She knew bronze-plated baby shoes existed, so somehow, somewhere, someone knew how to alchemize objects' exteriors. While she wasn't that someone, perhaps Doc was. After class, she walked up to his desk. She could already imagine the eyes of the other students, envious of the shiny anatomy augmenting her own like extra bionic ribs.

"This is what I want to do," she said, telling Doc about the bones and the baby shoes. "How do we do that?"

It's a simple and somewhat vain sentiment in this scenario. But "This is what I want to do. How do we do that?" is also the essence of science.

"I don't know," Doc said. "But I'm sure we can find out."

They scoured the library; Doc wrote letters to the shoe-coating companies, who told him the process was proprietary. But they gave enough hints for Doc to know he needed to combine acid and silver arsenate. And in a public school physics classroom, the two lowered the graphite-infused bone toward the chemical solution, turning it into an electroplating cell. Tarter and Doc watched as atoms attached themselves to the submerged bone. They lifted the object out of its bath, its crown breaching bright and promising. The next day, Tarter was the news item of the school day.

But later that week, she looked down to find the wishbone—covered in the scientifically shiniest metal on the planet—had already tarnished. She took it off and tried to scrub it clean but succeeded only in rubbing the shine itself off.

Today, when cough syrup sales are regulated and classroom dissections happen via software, this caustic-chemical collaboration would never be allowed. "Who knows what arsenic fumes were coming off that," she says now. "But Doc was just one of the best role models you could have about science, being curious, and figuring it out."

❖

The desire, perhaps even the need, to "figure it out" remains central to Tarter's personality today, when she is in her early seventies. As Tarter recalls aloud those high school years, she is sitting on the deck of her vacation house at Donner Lake in Northern California.

She still loves to be on the water, near large concentrations of the molecule that gives us, and possibly others beyond our atmosphere, life. (It's also the molecule that her husband, Jack Welch, discovered lives in the clouds of gas between the stars.) She and Welch bought this shoreline cabin in 1989. Decks protrude from each of the three stories, revealing a spectacular view of our planet on every level.

It is September 2014, and in the past hour Tarter has twice demonstrated the kind of inquiring mind Doc and her dad instilled. First, while chopping vegetables, she declared that she had always wanted to write a book called *Physics for Housewives*. "Just solutions to common problems," she said, "or how cooking with a convection oven alters your baking game plan, or how to fix a vacuum cleaner."

Second, when Welch declared that his electronic keyboard was out of tune, Tarter scavenged the Internet to find out how something electronic could possibly be sharp. This made no sense, since the notes came from frequency-generator chips, which are programmed to do only one thing—play the right note. "How can they be wrong?" she asked. It was not the "how" of incredulity, but the "how" of engineering—what physical mechanism could skew the tone?

She leaned over the edge of her easy chair, looking at the laptop planted on its footstool. An unexpected voltage spike could have reset the tuning away from the standard frequency, shifting all the keys, she learned.

The lake-facing living room wall, in front of her, is nearly all windows. Across from the wide glass panes, just inside the front door, wooden letters proclaim that this place is "Jill and Jack's" (and not, one will note, the reverse). A framed full text of Max Ehrmann's "Desiderata" hangs nearby, inscribed on lacquered parchment and backed with grainy wood. "You are a child of the universe no less than the trees and the stars; you have a right to be here," the early-twentieth-century poem says. "And whether or not it is clear to you, no doubt the universe is unfolding as it should."

Bookshelves play host to John Grisham thrillers, the *Cyclops Report* (a 1970s NASA report on SETI, still the field's scripture), a *Mathematica* manual, and a hardcover copy of *Vogue Sewing*. The room's centerpiece, though—the one that draws the eye—is a framed print of *Earthrise*. This iconic *Apollo* photo shows Earth hovering over the moon's horizon—the first view of our planet as a planet. It is Tarter's favorite astronomical image. Another copy of it hangs from the wall of her office at the SETI Institute, this one signed and dedicated by Bill Anders, the astronaut who took the photo while orbiting the moon on Christmas Day, 1968. When humans saw *Earthrise*, Tarter believes, we understood our cosmic connection for the first time: what exactly an Earth is, what it means that Earth is in this larger outer space.

As Tarter gathers her thoughts, preparing again to wander the halls of high school, she looks out over the lake, its water placid and cold. Donner is a high-altitude reservoir fed by snowmelt and streams both under- and aboveground. From the shore, the bottom drops out quickly. At its deepest point, it's 330 feet from surface to silt. She looks intently outward, like her teenage memories might lie on the other side of the lake. As if, as happens when we look out into the universe, she is seeing back in time.

"I wanted to take Shop," she says. "But no, I couldn't do that because I was a woman, so I had to take Home Economics."

She shrugs and looks through the sliding glass door at Jack, whose fingers meander over the sharp piano keys. "I learned to sew back then, and it's been incredibly useful," she says. "For a long time, I made all of Jack's shirts, because he loves paisley material and he's got a thin neck and ape-long arms."

She looks over at him, as if checking the evidence to make sure her conclusion is correct.

"And then I told them, 'I've taken Eome Ec. Now I want to take Shop.'"

❖

She was a tall teenager—taller than the boys, all the way through high school, even though she had skipped two grades. She wanted to compete athletically, but Title IX did not exist yet, so her school had only two options: cheerleader or drum majorette. "Clearly I was not going to be a cheerleader," she says. "Nobody was going to throw me up in the air. But, thunder thighs and all, I did in fact have the motor skills to do baton twirling."

This choreographed, costumed activity was a good thing, until Doc surprised Tarter by sponsoring her to take a physics course at Columbia University. College students, the big city: it would have been fantastic—except that the class met on Saturday mornings, when she had to be on the football field throwing shiny sticks into the air. The physics class stayed on the sidelines while Tarter spun batons with the centripetal force she had opted not to learn about.

"A really intellectual person would certainly have taken the physics opportunity," she says. "But I did it the same way I've always done things: the first commitment you make is the one you stick to."

You become an engineer, goddammit, and you go to drum major-ette practice.

While she was good with goals, she wasn't quite as adept with deadlines. As junior class treasurer—a campaign she'd won by giving out bags of jellybeans labeled "Vote for Jillybean"—she had volunteered to cover a school-prom arch in flowers. Flowers made of Kleenex. Tarter was tasked with folding, scrunching, stapling, and pinning all of these tissues before the event. But she had put it off, and instead of sitting down and scrunching, folding, stapling, and pinning, she was panicking.

Her mother finally asked her what was going on.

Tarter started crying and explained between gulps of air that she needed to make hundreds of paper-based plants . . . now.

Betty called her friends, mothers of Tarter's friends, and asked them all to come over and fold, scrunch, staple, and pin. "We're going to work until it's done," she told Tarter.

Her mother, here and in life in general, acted as the pragmatic complement to her father's American-dream-style encouragements. "When my dad was alive, and for a while afterward, he was the one who reinforced the idea that if you're willing to work hard enough, you can do anything," she says. "But what does that really mean? How do you 'do anything'? It took me a long time to realize that my mother gave me the skills I needed to actually do it."

Those skills included the ones necessary to pursue an engineering degree. Tarter applied to a number of New York colleges, but Cornell University especially caught her eye. After all, its first name was her last name. She and her mother did some detective work and discovered that Ezra Cornell, the school's founder, was her five-greats-back half-uncle—enough to qualify her, or so they thought, for a bloodline-based scholarship. Ezra himself may have had enough money to warrant a bronze statue in the middle of an ivy-invaded campus, but Tarter's family had never had much, especially not after her father died. So Betty composed a letter to the financial aid office, informing them of Tarter's genetic code and applying for the scholarship.

Cornell University wrote its great-great-great-great-great-half-niece. "Dear Mrs. Cornell," the reply read. "We are so sorry to inform you that this scholarship is only for male descendants."

Within a week, though, an unexpected envelope arrived, the address of Cornell's financial aid office staring from the upper left corner. Tearing it open, Tarter found a letter offering her full tuition and fees, courtesy of Procter & Gamble, for the five years usually required to get an engineering degree. They even agreed to let Tarter take the fifth year's money and use it during summer sessions so she could complete the program in four years.

❖

Once a year, executives from Procter & Gamble schlepped up to Ithaca to eat a meal with their scholarship students. Tarter was grateful for the free dinner, but the businessmen had no idea what to do with her, the only female student there. During Tarter's senior-year dress-up dinner, one of the suits said to her, "I hope you don't expect to have a job with us when you're finished."

Another of the professional engineers piped. "Actually, you know, we have this department that makes the chemical for home per-manents," he said. "You're enough of a woman to know what home permanents are, right?"

These men were far from the only ones whose antics she dealt with in college. While the engineering college's class of 1965 had 300 students, she was the only woman. The female dorms were all the way across campus from the engineering school. From November to April, Cornell's campus is a frozen wasteland of frostbite, split by icicled gorges. But even in single-digit weather, female students had to wear skirts once they crossed the all-too-symbolic bridge connecting their dorm area to the rest of the school. Pants were permitted only in their residence area.

She moved to campus a week early to participate in orientation, get settled in the dorm, and meet her all-male classmates. These boys fastened slide rules to their belts like nightsticks. The metal, inscribed with the tiny lines and integers that made the world a quantifiable thing, jangled unbecomingly against their thighs.

Tarter decided to date only architects. Architects, with their visions of an aesthetically pleasing world and their tools left safely on draft tables, were the opposite of engineers. Every two or three years, these architects took over an under-construction campus building. They transformed it into the kind of building *they* wanted and held a themed dance inside. Tarter, who began dating an architect her first year, was invited to attend.

That year, the theme was "Black Orpheus," an artsy reimagination of the legend of Orpheus and Eurydice, in which the former rescues the latter from the underworld. But this version was set in Rio de Janeiro.

The night before the dance, Tarter stayed up late fashioning her costume (having predictably put the project off). Holding a can of spray paint a few inches from a volleyball net, she pressed the button and plated it in aerosolized gold. The costume's immodesty meant she had to creatively cover herself, so in a cauldron constructed from a trash can and a hot plate, she created DIY body paint using lavender dye. Surveying herself in the mirror, she saw patches of purple skin peeking through each roped-off square. It was good. She scampered across the bridge toward her date and his party.

The pair propelled each other well around the dance floor, winning the evening's dance contest. She still has the prize: a stylized sun pendant, made of aluminum and black ebony wood. At the end of the night, the solar system's center laid against her chest, she returned to the dorms at 1 A.M., having received special permission to come back after curfew. But she had to wake up her dormmates to help her scrub the paint off her skin. "Everybody thought I was weird as heck," she says. "But they thought I was weird as heck anyway for being in engineering in the first place."

She was close enough to some of the female students to rouse them and take them into a shower stall. But she orbited, rather than inhabited, their world. She was a nerd. She worked all the time. And she worked alone, safely ensconced in pants on the cordoned-off part of campus, while the male engineering students split the homework and shared their solutions at the bottom of Libe Slope, right next to the engineering quad. Most of the girls in Tarter's dorm spent their evenings in the library, not to study but to socialize, and returned just in time for curfew with loud mouths and the latest gossip. She started going to bed at 10 P.M. and waking at 5 A.M., when the world was quiet and no one could watch her do vector analysis.

Tarter's dorm was next to Beebe Lake, a body of water flanked by Frisbee teams during the warmer months but frozen (like everything else) in the winter. Some days, Tarter finished her work early and glided across the ice, a rare frictionless space. The dorms, though, were locked until 6 A.M. She waited alone on the steps, ice skates dangling by knotted laces from her neck, pages of perfect equations behind her in a building of sleeping students. Still, while her peers didn't get her, they were mostly aloof, not often mean. There were exceptions, of course.

❖

The freshman Engineering Problems and Methods course, a 300-person lecture class, had homework problems like

A semitruck pulls over, and the driver gets out and bangs on the side. He does it again a few miles later. And again. And

39

again. When you can't stand it anymore, you pull off and ask him what's going on. "I've got five tons of pigeons on board, and my axle is only rated for two tons," he says, "so I have to keep the majority of the birds in the air at all times." Comment.

Tarter was excited to think critically, even if the topic was pigeons, even if it meant being surrounded by 299 boys. Betty was excited, too, and she wanted to take care of her only daughter even though she was away at school. Doing what she could, Betty ordered special cloth nametags and sewed them into all Tarter's clothes. She sat up late at night, her daughter asleep in another room, plunging the needle up and down and thinking with each prick through the fabric about her daughter. She didn't notice when the tag on one sweater declared the clothing to be "100% virgin wool."

One day in that beloved freshman course, Tarter shrugged that sweater off her shoulders. The boys behind her read the shirt's tag: "100% virgin Jill Cornell." An undercurrent of whispering swelled to a hum (Tarter, ever the physicist, characterizes the noise as "propagating backward and sideways"). The instructor stopped the lecture.

"What's going on?" he asked, peering out over the stadium seating. "What is this commotion?"

The boys looked back and forth at each other, then down at their laps, until one finally said, "Well, Jill's got a sweater that proclaims that she's one hundred percent virgin."

Tarter sat there silently, staring straight ahead and wishing a wormhole would open in the floor beneath her, ferrying her to a different part of the universe.

❖

Tarter's first year wasn't all embarrassment and isolation. After the architects, she began to date her physics teaching assistant, a grad student she'd met the first semester: C. Bruce Tarter. "However it happened and however appropriate or inappropriate it would be in the modern world, we were mutually attracted," says Bruce.

Bruce was smart. Bruce was motivated. And, Tarter thought, Bruce was worldly and sophisticated. He'd grown up as a Southern gentleman—a society man from Louisville, Kentucky. Such an upbringing wowed Tarter, who was far from a debutante.

Bruce shared an off-campus apartment in an old house with another physics graduate student, named Hakki Ögelman. Ögelman had a girlfriend, too: Ivy. The foursome often ate dinner together. At one dinner, Ögelman, who later became a pipe-smoking gamma-ray astronomer, attempted to explain to Ivy why salt thrown onto the gas stovetop turned the flame yellow. When they meet fire, the electrons inside the sodium atom jump from a high energy level to a lower one, letting loose a single photon of light when they do so. This photon has a specific wavelength, and every visible wavelength of light represents a specific color. For scorched sodium, that color is a nearly green yellow of 589.3 nanometers.

But this is hard to explain in words, especially for a man who'd been shooting bourbon all night. Instead of talking about 3p and 3s atomic energy shells, Ögelman jumped on and off a kitchen chair—high energy, low energy, photon; high energy, low energy, photon—until the chair broke.

❖

That summer, as he had the summer before, Bruce got an internship at Lawrence Livermore Labs in California. Tarter, just 18, accompanied him. "Her mother was violently opposed, in the mores of that time, so we found a friend to bribe to 'travel' with us," says Bruce.

For the first months, Tarter got an apartment in Berkeley and took summer classes at the university. When she finished her courses, she went out to Livermore to live with Bruce and worked at a Chinese restaurant. Together, they settled into a California life. Their new world could never be silenced by snow. The Pacific was always nearby, throwing itself cold against the rocky coastline. People were looser, happier than in the Northeast. It was then that Tarter decided she wanted end up there for the rest of her life. She made the decision in the same way she made all big life decisions—quickly, with finality. Goddammit.

Once she returned to Cornell, Tarter didn't want to wave to the dorm mother when she walked in at 9:59 P.M. intoxicated, from both liquor and her receding sexual horizons. She didn't want to walk in at 9:59 P.M., period.

❖

Upon their return to Cornell, during her junior year, Tarter and Bruce decided to marry at the faculty club. Bruce's parents loved Jill, but in photo albums, Jill's mother, Betty, shows up gritting her teeth in a nice dress. She was not a fan of the union. Procter & Gamble wasn't either. If Tarter wanted a husband, it was clear she didn't want a degree and wasn't a serious enough student to deserve their money. The company retracted her scholarship. Enraged, she marched to Dean Dale Corson's office.

"We were counting on that money," she said to him. She and Bruce had run the numbers for upcoming grad school finances—and certainly hadn't made any mathematical mistakes—and they wouldn't be able to swing it without this scholarship.

"I just don't think this is fair," she concluded.

She stood there, arms taut like baseball bats.

Corson put his hands together and nodded. "I agree," he said.

He stepped up to the plate in her defense. Someone like Tarter was not going to drop off the face of the academic Earth, he told the company, just because someone bought her a nice ring. It was a home run.

Years later, Corson became Cornell's president, at the same time that Tarter was on the President's Advisory Committee for the Arecibo radio telescope, which the university then operated for the National Science Foundation. Tarter walked up to President Corson at one of the meetings. "See? I didn't just quit and have babies," she told him. "I'm a PhD. I'm your advisor on this observatory you're running. Thank you so much."

Tarter finished the engineering program in four years, with a GPA that should have gotten her inducted into the engineering society Tau Beta Pi. But when she applied, she was told she could not be a full member because she was a woman. Years later, when the gender policy changed, a Tau Beta Pi representative called to say she could now join, for a few thousand dollars.

"It hasn't done me a damn bit of good so far, and it isn't going to," she said. "Enjoy your day."

❖

After Tarter graduated, she didn't go to work test-tubing perm chemicals. She didn't want to be any kind of engineer at all, especially not one sweating under a lab hood for beauty products. She hated the slide rules, and strict rules in general. She was going to find something else to do. Now that she had graduated, she felt she had achieved her goal of "becoming" an engineer but needed a new path for the rest of her life.

She stuck around Ithaca, waiting for Bruce to finish his PhD, which took him a year longer than expected. But rather than sitting idle, she sampled classes in other departments—Russian, world history, nuclear physics, and astronomy. Astronomer Ed Salpeter's class dealt with star formation: *How* do stars become stars? *Why* are there stars in the first place?

These, she thought, were interesting problems to solve. She applied to graduate school in astronomy. She would become an astronomer, goddammit.

❖

Her tendency to jump in hasn't changed in the five decades since Salpeter's astronomy course. It is 8 A.M. at the cabin on Donner Lake, and early-morning fog hovers above the water. The air temperature is holding in the 50s. The lake water, still in shadow, nudges the mercury to just 62 degrees. Nevertheless, a bathing-suited figure stands on the edge of the dock below "Jill and Jack's." The silhouette dives into air-clear water.

"Not too cold," Tarter says when she pops back up, surveying the landscape. Jack is still asleep. Mountains sprout from the lake's edges. With their Douglas firs and vocal tree squirrels, they seem habitable. But in a couple of months, they'll be covered in the kind of head-deep snow that left the pioneering Donner Party stranded. Trapped in a six-month-long, 20-foot snowpack, many died. As you likely learned in some horrifying history class, the remaining party members ate the bodies.

When asked if she would become a cannibal in such a circumstance, Tarter smiles before she says, "Yes, absolutely."

CHAPTER 2

BABIES, BROWN DWARFS, AND BIG MOVES

Physicist Hans Bethe stood before his quantum mechanics class, fingers holding the nub of chalk against the upper left-hand corner of the blackboard. By the end of the lecture, he would fill that board plus another below and two next to it, sticking a perfect landing at the lower right-hand corner. The length of his notes was a known quantity, although the point of quantum mechanics is that the universe is, at its most fundamental level, uncertain.

Tarter had recently graduated from Cornell and was waiting for Bruce to finish his degree. She had chosen this extra class because she adored Bethe. Earlier in his career, he had helped mushroom atomic theory as a member of the Manhattan Project, although he campaigned against the bomb itself. In this atomic research, though, he realized something important: the same ideas that allow humans to rip individual atoms apart, leaving trees bare and genes skewed, also play out 93 million miles away. Inside the sun—and inside all other stars—atoms combine to create new elements and release the warmth and light that bathe our planet. This discovery won Bethe the Nobel Prize in 1967.

Tarter copied down his chalky scrawling, feeling only intermittent twinges from her torso. This was a good day. Some days, she didn't make it to class at all. The week before, as she walked toward the large lecture room, with its heavy uncomfortable oak chairs, she felt the nausea rising again, like a fist thrusting up her esophagus. At the last second, she ran into the bathroom across from the lecture hall. As much as she loved Bethe and Bose-Einstein distributions, morning sickness won.

This was not what she'd had planned when she and Bruce decided to get pregnant. If humans were supposed to have babies, why did the body act like the fetus was something foreign to be fought? But she was in it now, the cells splitting like atoms in a chain reaction that would last six more months. She couldn't say, "No, this isn't what I asked for." And it wouldn't kill her. Difficult experiences are difficult because they make you live through them.

Of that time, generally, and their relationship then, Bruce says only, "The pressures were never minor."

❖

During those in-between months in Ithaca, when Tarter was sick and sampling classes and Bruce was finishing his doctorate, she also entered the social life of the Cornell astronomers.

At one department party, of which there were many subsequent and similar iterations, the smell of the Sara Lee pound cake crept into the Salpeters' living room, where Tarter was dancing. Everyone else in the department had long ago stopped, and were probably either amused or embarrassed by her exuberance. These others moved toward the kitchen and conversation, but Tarter wasn't finished moving and shaking. Only after she couldn't dance any more did she walk to the Salpeters' kitchen. People crowded around the counter. Tarter wiped sweat away and joined its periphery, watching people pick particles from their pound cake. Miriam (Mika) Salpeter, Ed's wife, had set a number of the desserts out to thaw hours earlier and warmed them in succession, supplying them as demand warranted.

Mika was a prototypical hostess, but not in the singular way of most faculty wives. She had a doctorate in psychology and was interested in neuroanatomy and neuromuscular disorders. Normally, Tarter thought being a faculty wife was "a fate worse than death," even when the wife had serious academic credentials. The university did not often hire female spouses. According to a memorial state-ment Cornell put out after Mika's death, after her only male ally in the department where she was a postdoc left, "her chances for an academic post at Cornell were reduced to nil. There was downright disbelief at the time that academic performance could be combined with motherhood, and Mika did not initially escape the consequences of such misjudgment."

Many women's careers ended up on ice. But Mika continued her research, as a non-faculty member, investigating the neurons that control voluntary movement. She became a renowned expert on the synapse, writing seminal research papers. Still, Cornell didn't hire Mika as faculty until 1967, 16 years after she and Ed moved perma-nently to Ithaca and long after she was an established scientist. She, said Cornell's statement, "finally receiv[ed] the professorship she deserved . . . [and] became a strong role model and rights advocate both at Cornell, and nationally within her professional community."

Mika, even at the time, expanded the possibilities of what "female scientist" meant to Tarter. "I was looking for role models," Tarter says, "always looking for role models. You can be good at what you do in science. And yet you can be the social heart."

At the party, Mika stood next to her pound cake and told jokes and anecdotes. "She was a master raconteur, who told stories to diffuse tension, or simply to bring joy," her colleagues wrote after her death, "and she used this talent with enormous success as chair, colleague, and friend."

But then she could launch just as powerfully into the cellular causes of neuromuscular disorders. "Whether over the kitchen table or on the ski lifts, she never hesitated to bring up science," continues the remembrance. "The intellectual exchanges between her and Ed were exciting and memorable to participants."

Later in life, Mika and Ed together created a technique to see individual signaling molecules in cells, the chemicals that tell our bodies what to do. The duo mapped our bodies' smallest constituents, just as Bethe dug down into an elemental part of stars. Their science was about getting to the core of things.

But that night at the party, Tarter couldn't look away from how Ed and Mika jousted and joked intellectually. It was an equilibrium she wasn't used to seeing among spouses.

Tarter reached for the pound cake. Leaning back, she chewed, her body still radiating the heat it had generated.

❖

After Bruce finally finished his degree, NASA's go-to communications vendor wooed him with a large salary. He knew he didn't want to be an academic, and so, out of his many job offers, he'd chosen this one—the most industrial one he could find. "In those days if you could sign your name, you could get a reasonable job," says Bruce. The company, Aeronutronic (an Orwellian portmanteau of *aeronautics*, *nucleonics*, and *electronics*) was in Newport Beach, California. Bruce drove their car across the country to their new home, while Tarter (still pregnant) flew with their cat, Petronius, who was named after the feline in Robert Heinlein's sci-fi novel *The Door into Summer*. The fictional cat refused to leave the house if there was snow on the ground. Instead, he trekked from door to door, sure that one of them must be "the door into summer." He believed that by changing his location, he could change his circumstances.

Newport Beach was full of doors into summer. The town sits along the last coastal curve before the Mexican border, north of San Diego and south of Los Angeles. It's a liminal place, isolated inside its own surf-, beer-, and boat-centered existence. It boasts the only airport named after an actor (John Wayne) and the most dangerous surfing spot in the United States.

At this spot, known as the Wedge, when a wave approaches from the south, it hits a rock jetty constructed in the 1930s. That collision sends a reflected wave back toward open ocean. The backward-traveling wave collides with the next incoming wave, and they add together into a larger wave. *That* wave then hits the jetty and sends yet *another* reflected wave back—you see the pattern. The process happens fast and progresses chaotically, in the mathematical sense: small changes in the starting conditions lead to radically different and sometimes catastrophic outcomes—out-of-nowhere 30-foot swells, riptides that want to drag you to the edge of Earth. Even the most weathered surfer can't always predict the waves' breaking points, which are often in only ankle-deep water. It doesn't take much imagination to picture the bones that could snap. Engineers built the jetty to give Newport a harbor—a calm, safe space. But they also inadvertently created a monster. Surfers there say, "Small days you'll break your back, and big days you'll drown."

Newport Beach is largely the same now as it was in the 1960s, aside from the price of real estate and the disappearance of pastel hotel signage. An ocean inlet cuts into the shoreline, creating two islands that look like the lenses in a pair of sunglasses. Every piece of land has long slips reaching out. The water channel snakes inland, where it crosses the Pacific Coast Highway. This highway, Tarter often thought as she and Bruce sat in traffic, was the only path to the only hospital. And it was usually packed with tourists in polygonal Art Deco cars and Volkswagen vans.

Luckily, their daughter, Shana, decided to emerge into the world in the middle of a July night, when traffic was light. Bruce and Tarter sailed to the hospital, and soon Shana emerged screaming onto this planet.

Another being. It was so simple, so natural, like some fundamental law of the universe.

❖

Tarter planned to start grad school at UC Irvine, just inland from Newport, the fall after Shana was born. Bruce's job was great, good-paying, solid. They had a nice apartment. They were secure. Those conditions could easily have led to a clear-cut, SoCal, all-American family. Then one day, not long after they moved, Bruce told Tarter they needed to talk: he didn't want to work at Aeronutronic anymore. So he wrote to Lawrence Livermore National Laboratory, just east of San Francisco, where he'd done his summer internships. And soon, they sent him another job offer (he is currently the lab's director).

"We're putting graduate school on hold," Bruce told her.

Still, despite this tension, Bruce's dissatisfaction with his career, and Tarter's own limbo, they both have fond memories of their brief stint in SoCal. "I remember it as one of the better times," says Bruce.

❖

They moved up to an apartment in Danville, next to manatee-like hills, that summer, with canyons that glowed gold in the summer and green in the winter. The seasonal grass there shoots up after the first rain, more mushroom than plant. Poison oak, too, grew in this climate. Tarter always had a rash blooming across her forearms, right where the cat rubbed against her. She took to sipping the toxic urushiol oil in dilution, thinking a confrontation with the poison in miniature might help, a remedy much touted by the rangers who worked in the canyons behind their home.

The fall application deadline for nearby UC Berkeley had already passed, so Tarter stayed home to take care of Shana, who was starting to speak. When they walked to the grocery store, Shana pointed to ads in the window and said words aloud. Tarter told herself Shana couldn't be reading already—could she? Shana then tried to prevent Tarter from doing the same, by disposing of Tarter's glasses down a storm drain and pulling all the books from their shelves.

When this year was over, Tarter was more than ready to accept UC Berkeley's offer of admission to their graduate astronomy program.

❖

On the first day of graduate school, Tarter walked toward Campbell Hall on Berkeley's campus, on her way to see the chair of the astronomy department, Ivan King. The building was a plain rectangle with a flat roof broken by a telescope dome. Grids of perfectly spaced windows peered from all four sides. Tarter stepped through the doors, uncertain how things would go but sure this was the right step. Here she was, in Berkeley, about to find out why stars exist in the first place. Although waiting for Bruce to finish school, waiting to get pregnant, and leaving Newport Beach were not part of the plan, they had led her, eventually, here, to the endpoint she'd always envisioned, beginning with those nights on the beach.

She arrived in King's office. He had piled academic journals on every surface—monoliths and ziggurats honoring the gods of peer review. The *Astrophysical Journal*'s 1963 editions; *Nature* volume 204, issue 4965. If you asked him to find the seminal article about the initial mass function of dwarf galaxies, he would reach one-sixth of the way down the stack next to the coatrack and whip out the right issue. In an interview with the American Institute of Physics, he claimed that a youthful administration of the IQ test gave him a 196. "I gather it's sufficiently off-scale that a number like that is not terribly significant," he said to David DeVorkin, "but that was the thing that hit the newspapers with a splash."

They greeted each other, and the two other female graduate students—Kate Brooks and Linda Schweizer—soon sat down with them.

"You three ladies are so lucky," King began, "that all the smart men got drafted for Vietnam."

In other words, spaces existed for them in the department only because more-brilliant boys were away at war. Tarter went rigid, blood rushing toward the core of her. Although she didn't look at the other women, she sensed the same process was taking place inside of them—fight or flight. But they all battled their animal instincts, while King smiled unknowingly at them. When the orienting conversation was over, Tarter, Brooks, and Schweizer walked into the hallway and watched King's door shut behind them.

"What the fuck?" they repeated to each other, knowing there wasn't a satisfactory answer.

The science Tarter saw around her in the coming years at Berkeley—the observational kind King did—wasn't the kind that appealed to her. "I wanted to know how it worked," she says. "The old model of going up to the telescope on the mountain night after night, making images and making images—it wasn't my thing." She wanted to bring a bit of the engineer in her to astronomy.

It was a seed of discontent that would eventually grow into a SETI career.

❖

During that first year at Berkeley, professor Len Kuhi hired Tarter to program a telescopic instrument called a spectrometer. Spectrometers parse the light—which scientists sometimes call "electromagnetic radiation"—that comes into a telescope. Light is the only way we can learn about the universe. We'll never touch a star or feel a black hole's gravitational field, but we can collect light (photons) from them. Although they have no mass—photons are, materially, nothing—they are our only connection to the cosmos, and they come in lots of different sizes and energies. The highest-energy photons are called gamma rays, and the lowest energy photons are called decametric radio waves. The visible light that our eyes sense is middling in energy.

All of this light travels in waves. The length of each wave—the distance between each crest—represents a different color. A wave with a length of 500.1 nanometers is slightly bluer than a wave of 500.2 nanometers. Your eyes can't catch that subtlety. But astronomers build their tools better than the eye, to give humans sensory superpowers. Kuhi's spectrometer could spot the small differences between light waves, separate them according to their lengths, and help interpret what those differences meant. Light of 589 nanometers indicates sodium in a star, while a brightness that peaks at 883 nanometers tells you that the star is 9,980 degrees Fahrenheit.

Berkeley had a 30-inch optical telescope, and Kuhi had hooked the spectrometer to it like a Lego. Tarter's job was to automate the spectrometer so that even hapless students could use it. Kuhi showed her the PDP 8/S computer she would use for the work. It was the first personal "minicomputer" (a svelte 100 pounds), and the first computer at all to cost less than $20,000 (which, at the time, would have bought a nice house in the suburbs). They called it a desktop because it technically could sit on the top of a desk, but it hardly resembles those of today. It had no language, experiencing instead permanent aphasia: the user had to speak in symbols—long strings of 1s and 0s, written in an unintuitive base-8 code called octal. And even with this simple code, the computer could understand just 11 commands, which the user had to string together in combinations. A set of rust- and bone-colored switches sat in the console, above a panel of lights with inscrutables next to them: TAD, ISZ, DCA, ION, PARITY.

She punched her octal onto one-inch gray tape, which she had to unspool from long rolls. When she made a mistake, she taped over the mispunched holes and poked through where she should have put holes in the first place. The patchwork told the tale of every error.

She rolled the tape back up so the PDP 8/S could suck it in and read it. It was awful.

When the project finished, she bid farewell to the PDP 8/S, hoping never to see it or its parity errors ever again. But the dinosaur technology would lead her to SETI, which then led to her becoming Jill Tarter, alien hunter.

❖

But she still wasn't there yet. She was becoming just a regular radio astronomer, and so she attended her first American Astronomical Society meeting in San Juan, Puerto Rico, in 1971.

"I don't think either she or I had ever been out of the continental United States at the time," says Bruce. On top of that personal specialness, Bruce—at least retrospectively—sees the meeting as a watershed moment for astronomy in general and young astronomers, like them, in particular. "Astronomy was essentially a fairly moribund, uninteresting field up until the mid-1960s," he says. It was like early biology, where people just cataloged and classified (although in the case of astronomy, they were classifying stars and not jungle birds). "In the 1960s, it exploded," he continues. "There were black holes, starting in 1963, and there were X-ray sources. The entire field had just gone off like a supernova."

A bunch of physicists, like Bruce, became *astro*physicists. And all of that excitement about the novel and energetic universe, and all the energetic and novel people, came together at the meeting. There was astronomy going on, sure. But also Jeep rides through the jungle, casinos, booze, drugs, sex. "The world was just turning on," says Bruce, "and the social revolution was there. You put that together with a lot of young people, and . . ." He trails off, and specifies that he didn't do the drugs—he had a security clearance to keep.

After the official conference and its unofficial debauchery, Tarter and Bruce didn't have to fly immediately back to California.

"Come on," said Stuart Bowyer, a Berkeley professor, said to them, gesturing in the general direction of the ocean, which was visible between the swarm of hotels along Condado Beach. "I'm going to rent a surfboard."

It was so hot and humid that they couldn't tell if they were sweating or if the air was just condensing on their skin. The ocean sounded perfect to Tarter. Bruce disagreed, preferring to stay in the hotel room and work, so Tarter went on her own. When Bruce tells the story today, a bit of old jealousy creeps into his voice.

At Condado Beach, the waves broke tepidly, more rolling than slamming into shore. On the surfboard, Tarter first felt good, balanced, aware of the ends of her limbs. But then her feet began to lose their place. She didn't know where she stood, couldn't tell where her body was in space. She slipped toward the water.

Just beneath the surface, thousands of individual polyps had been secreting calcium carbonate for years. This substance hardened, bit by bit, to form a coral reef.

When Tarter's thigh collided with the coral, it slashed big gouges into her leg. The crash broke some polyps off and sent them drifting out to sea; others remained stuck in her leg.

When Tarter returned to the hotel lobby, blood everywhere, Bruce took one look at her and headed to the bar.

"You got yourself into this," he told her. "You can get yourself out."

Today, he says, "I think the Puerto Rico thing was—if not an epiphany—probably in a funny way the time when Jill and I began to realize that our paths had diverged."

❖

During Tarter's second year of graduate school, in 1969, she became a teaching assistant for surfing Stuart Bowyer's general astronomy course. He used a textbook called *Intelligent Life in the Universe*, written by Ivan Shklovskii and Carl Sagan. Shklovskii wrote the tome in Russian and allowed Sagan to edit the American edition.

Sagan "edited" liberally, doubling the book's length to broaden both its subject matter and audience. It was Tarter's first introduction to academic study of the possibility of intelligent life in the cosmos.

The book begins with the beginnings, in a section about cosmology, talking about "the universe and its parts," from stars and galaxies to how they got there. In the second section, it looks into the origins of life on Earth—which we don't fully understand now and certainly didn't in 1966! But the authors speculated, in this chapter, about whether biology could have arisen on other planets. The third section takes the questions a step further, speculating about whether any of that hypothetical biology could have evolved into something smart, something whose technology we could find. It looks at radio searches for such life, like the one Frank Drake had performed just a few years earlier, and searches for laser signals, which scientists would not undertake for decades. And although Tarter enjoyed discussing it with the students, it didn't tell her her future. "It is a bit ironic that this first brush with SETI left me pretty untouched," she says. "I never would have predicted that Carl Sagan would become a colleague and SETI would be my focus."

Bowyer looked like Buddy Holly and was prone to making grand statements during lecture. The students brought these statements to Tarter in their homework sessions. "What did Professor Bowyer mean?" they asked.

She couldn't substantiate the claims. And worse, she wasn't exactly prepared to answer their more grounded questions, and once accidentally drew a plot with an axis labeled as "wavelength" when it should have read "frequency," teaching them that hot objects mostly emit photons at lower wavelengths, when the opposite is true. Her knowledge ran a bit ahead of theirs, but not by much, and her poor end-of-year student evaluations reflected that. A tribunal of faculty members, who sat stern behind a desk, interrogated her about her experience at the end of the course. They told her that, under other circumstances, they would help her improve her teaching skills. But because Bruce had a high-paying science job, and they had limited money, they had decided to give her teaching fellowship to someone else.

"If I had done a brilliant job," Tarter says, "I could have sat there on my laurels and said I was a good teaching assistant. But I wasn't such a great teaching assistant. I didn't prepare enough. So I couldn't say, 'Screw you. This is really unfair.'"

Helping the youth of America with their homework just hadn't topped her priority list. She was wrapped up in her own academic work, thinking about what her research topic might be. She was now working with astronomer Joe Silk, looking at the ways gas is stripped from spiral galaxies. She thought the research might become her doctoral thesis. But she first had to finish her general courses and take a comprehensive oral test, called a prelim. And then there was also her daughter, Shana. Shana wasn't even in preschool yet. Every day, Tarter dropped her off at the babysitter and picked her up on the way home. That was anomalous, Bruce notes, at the time: no one sent their kids to daycare.

When Shana reached elementary school, the bus would drop her off at the Berkeley city library after school. After a few hours of reading and homework she would make her way to Tarter's UC Berkeley building. If Shana arrived after 5, the door was locked, and she simply waited outside doing cartwheels until her mother realized it was 5:10 and her daughter wasn't bouncing up and down next to her desk (or until a colleague came by to say, "There's some little girl doing cartwheels on the stairs").

Shana expresses no bad memories of this time period, no sense that her parents weren't paying attention to her. She does recall, throughout childhood, being surrounded by scientists. "I have memories of falling asleep on scientists' beds during parties," she says. "Their dinner-table conversation was rich with the day's work, the week's work, and drama. Some of it's good; some of it, you ignore."

❖

When it came time for Tarter to take her prelim exam, the professors grilled her for four hours. But when she passed, she did a cartwheel in the hallway. Trees, it turns out, don't fall far from apples.

And also, more conventionally, vice versa. From her earliest years, Shana could often be found lying atop a bear rug at the Danville house, wearing clothes Tarter had made. She stretched her arms along its arms and smiled like her teeth were as big as its teeth. This rug was an heirloom, of sorts. When Tarter was four years old, her father had gone out hunting for a deer and had returned instead with a black bear.

The family didn't have a big enough freezer for its body, so Dick prepared it, hung it in the garage, and went out to rent space in someone else's freezer. Alone with the major ursa, Tarter watched the blood drip from its nose into a pool of plasma on the garage floor.

She charged her friends a penny to come see it.

Later, the bear (its meat spoiled) became this rug. When Tarter needed comfort, as a child, she would lie on top of it and wrap its paws around her shoulders, its huge face above hers turning her into a two-headed alien creature.

Shana, the other bear lover, went on to become the assistant director at the Wilderness Medicine Institute in Lander, Wyoming.

❖

While Tarter was becoming an engineer, then a mother, and then an astronomer, SETI was busy gaining momentum as a field. The modern SETI movement began with the coincident timing of Frank Drake's first experiment in Green Bank—pointing radio telescopes at two sun-like stars—and a journal article by astronomers Philip Morrison and Giuseppe Cocconi.

Morrison had just finished work on the Manhattan Project when he took a professorship at Cornell University. One spring day in 1959, Cocconi came to his office and struck up a conversation about gamma rays and how the universe makes them.

"We realized we knew how to make them, too," Morrison said in a 1990 interview for the book *SETI Pioneers*. "We were making lots of them downstairs at the Cornell synchrotron." At that particle accelerator, scientists whirred electrons up to high speeds. These negative particles whizzed around a huge circular tube, emitting high-energy radiation—gamma rays. If particle accelerators could make gamma rays, then humans could harness them and beam them out into the universe. Gamma rays, because they have so much energy, travel straight through obstacles in the same way that X-rays go straight through your skin and reveal your bones. Because these rays are basically unstoppable, humans could use them to communicate across interstellar distances. And, perhaps, so could other species on other planets.

But Morrison thought they should not just accept that gamma rays would make the best missive. They should look at the whole electromagnetic spectrum to find the best wavelength. X-rays, visible light, infrared radiation, radio waves—what about them?

Gamma rays are great, but making them requires a lot of energy. No one knew how to create ultraviolet radiation. Dust particles in space scatter and absorb visible light—which has a medium-sized wavelength, a middling amount of energy. But long-wavelength, low-energy radio waves: We could make those so easily that we used them to play music in our houses, catch spy planes, and talk to *each other*. Why not use them to talk to extraterrestrials? And, by extrapolation, why wouldn't extraterrestrials use them to talk to us? (Looking for radio communications *from* aliens wasn't a brand-new idea: in 1924, when Mars slid close past us, radio operators trained their antennas on the Red Planet to see if any Martians were talking.)

Morrison and Cocconi decided that radio waves represented the best form of interstellar communication. But *which* radio waves? Radio waves can have wavelengths ranging from 1 millimeter to 100 kilometers.

Within that wide swath, hydrogen atoms—which make up 74 percent of the universe—emit radio waves that are precisely 21 centimeters long. Cocconi and Morrison thought any astronomically competent society would have discovered hydrogen's radio waves. And any halfway intelligent species would have built instruments specifically to detect those waves. And they might think that another halfway intelligent species (like humans) would have done so, too. Cosmic communication conceivably might concentrate around this wavelength.

The two scientists wrote up their ideas in a two-page paper called "Searching for Interstellar Communications," which appeared in the September 19, 1959, issue of *Nature*. "The reader may seek to consign these speculations wholly to the domain of science fiction," they wrote. "We submit, rather, that the forgoing line of argument demonstrates that the presence of interstellar signals is entirely consistent with all we now know, and that if signals are present the means of detecting them is now at hand."

And then came this famous line, still a SETI rallying cry: "The probability of success is difficult to estimate, but if we never search, the chance of success is zero."

❖

When Morrison and Cocconi's article came out, the young Frank Drake had already been planning his Green Bank experiment for six months. It even had a name: Project Ozma, after the strange land in L. Frank Baum's Oz novels. Drake had a longstanding and silent interest in extraterrestrial intelligence, heightened by an experience he'd had in graduate school. Late one night while observing radio waves from the Pleiades, he came across a "narrowband" signal (one that appears at a single frequency, like our radio stations) that looked "intelligent."

"What I felt was not a normal emotion," Drake told William Poundstone, Carl Sagan's biographer. "It was probably the sensation people have when they see what to them is a miracle: You know that the world is going to be quite a different place—and you are the only one who knows."

But when Drake tilted the telescope away from the Pleiades, the signal continued. Because it remained when the star cluster was out of the telescope's sight, it couldn't be coming from the cluster: It had to come from Earth. Still, the possibility of that miracle's reality stayed with him. After that, whenever he thought about using a particular telescope he would ask himself, "Could this be used to search for intelligent life?"

"The answer was always, 'No,'" he said in the book *SETI Pioneers*, "until we came to modern radio telescopes."

Drake went to work with these modern radio telescopes, at the National Radio Astronomy Observatory in Green Bank, West Virginia. Fifty miles from a grocery store, the Green Bank observatory is a scientific oasis in the middle of the Monongahela National Forest. It was a brand-new facility when Drake was contemplating his first experiment, cobbled together from bought-up farmland. Engineers and technicians grew radio telescopes from the ground up. They lived in pop-up construction-worker shacks and the eminent domain of farmhouses. According to Drake's calculations, the first telescope they finished at the site—the 85-foot-wide Tatel Telescope—could detect broadcasts as strong as the ones humans then made, if those broadcasts came from star systems up to 10 light-years away.

A few times a week, a group of Green Bank scientists would have lunch at a diner (which we might as well call "*the* diner," as it was the only one, but which they jokingly called Pierre's). And one afternoon, as snow fell on the only highway in or out of town, Drake told them about his conclusion: they could do a search for extraterrestrial intelligence right then and there. Lloyd Berkner, then director of the observatory, stuck a fry in his mouth and took a sip of soda.

"I like it," he said.

The Green Bank crew was isolated not just from fresh produce but also from outside ideas about SETI. But really, no one talked about it, anyway. "The subject was not discussed," Drake continued in the interview. "There was no way to know who was interested, just no way to make contact or to learn of other people's interests."

So the Green Bank scientists set about silently making their equipment and planning their experiment, keeping it quiet so neither their colleagues nor the press would descend. And then they saw the Cocconi and Morrison paper, laying out how and why to do a search exactly like the one they were planning. By then, astronomer Otto Struve had become the observatory's director, and he wanted his institution to receive credit for its alien ideas. He went public with the project at a lecture he gave at MIT a month later.

Drake quietly continued his work. Each morning of the experiment, he had to climb up to the garbage-can-sized instrument—an amplifier, which turns up the volume on signals from space—hanging above the telescope's dish. It required adjustment every morning.

"When I think of Project Ozma," he said in *SETI Pioneers*, "I recall how cold it is at Green Bank at four in the morning."

❖

Drake's project showed up in the pages of *Time*, where Hewlett-Packard executive Bernard Oliver saw it. He became a little obsessed. The next time Oliver visited Washington, DC, just a few hours from Green Bank, he called Drake at the observatory, an event Drake related in *SETI Pioneers*.

"Would it be possible to visit you and see what the apparatus is you're using in Ozma?" he asked.

"I'd be very happy to have you," Drake responded, "but you can't get from where you are and back in a day."

"Don't underestimate me," Oliver said.

The next morning, he flew in on a small plane and landed on the tiny airstrip next to the big telescope. Oliver's chutzpah, his electronics wizardry, and his fixation on alien intelligence netted him an invitation to the SETI meeting of the century, one called by J. Peter Pearman of the National Academy of Sciences.

❖

Now that the idea of ET was out in the world, Pearman wanted to build federal support for it. The best way to do that, he decided, was to convene a scientific conference on the topic, held where the whole thing started—Green Bank, conveniently a town where no one would notice a bunch of eggheads talking about aliens. He presented the meeting idea to Drake, and together they drafted a list of invitees.

Now, and to some extent then, it reads like a who's who of the scientific frontier: Carl Sagan, just 26 and almost an interloper; Melvin Calvin, who won a Nobel Prize in Chemistry during the meeting; John Lilly, a researcher who wanted to communicate with dolphins; Morrison; Oliver; and a few others. The meeting took place in the dormitory lounge at the Green Bank observatory.

To begin, Drake stepped up to the chalkboard and scrawled an equation. He didn't call it the Drake equation, but that's what scientists have called it ever since. It listed the seven factors that lead to the development (or not) of communicative, intelligent life. If they could figure out these factors, Drake said, and multiply them together, they would know how many civilizations populated the galaxy.

- How often do life-friendly stars form?
- What fraction of those stars host planets?
- On how many of each star's planets could life live?
- What fraction of those planets actually develop life?
- What fraction of that life evolves the kind of intelligence we would call intelligence?
- What fraction of that intelligent life communicates across the cosmos?
- And how long do those communicative civilizations last?

Drake turned around to face his distinguished audience. So?

They discussed the numbers for three days, coming up with a final estimate of "between 1,000 and 100,000,000" communicative civilizations in the galaxy. Because their talk occasionally diverted into dolphin territory—discussions of whether dolphins counted as intelligent (meaning intelligence arose at least twice on Earth—three times if we count what we now know about the consciousness of octopi!) and, more importantly, whether dolphins could get erections on command—the group was dubbed, after the fact, the Order of the Dolphin.

❖

In October 2010, the Green Bank observatory had a conference— Ozma@50—to commemorate the 50th anniversary of Drake's earliest SETI work and that first conference. The observatory looked much as it did in Drake's day, with the notable addition of the white-paneled Green Bank Telescope, taller than the Statue of Liberty and wide enough to cradle 2.5 football fields. But the town hasn't grown. Its population sign reads 143. Even elementary schoolchildren get the first few days of hunting season off from school. Tourists—often in caravans of motorcycles—driving along the one-way-in-one-way-out US 250 get bucolic views: barns, mountain ridges, analog gas pumps. Then, when they maneuver through a bend in the road, they find hulking technological structures plopped in this middle of nowhere. Next to these high-tech telescopes stand the buildings from the 1950s and 60s, which have been kept in a historical icebox.

Drake returned to town for the Ozma@50 conference, as did 48 others, including Tarter's husband, Jack Welch, a radio engineer. For the opening reception, guests gathered in the same low-ceilinged room where Drake wrote the Drake equation before it became the Drake equation. Now, a plaque with this formula hangs humbly over the fireplace. Aside from that, the space is the same. The same cheesy chandelier, reminiscent of a lava lamp, still hangs over the table. Orange vinyl couches and flat turquoise carpet still vie for your eyes' attention. A hinged chalkboard might actually be the same one Drake wrote on. In SETI, it seems, not much has changed.

❖

Barney Oliver, of the rogue plane flight into Green Bank, became a SETI advocate, speaking publicly whenever he got—or gave himself—the chance. NASA decided to put his general's voice to use: they asked him to chair a committee commissioned to think about "what would be required in hardware, manpower, time, and funding to mount a realistic effort, using present (or near-term future) state-of-the-art techniques, aimed at detecting the existence of extraterrestrial (extrasolar system) intelligent life." The committee met over the course of 10 weeks, dreaming up the specifics of a SETI scientist's dream telescope. What technological beast could best see—or hear—fantastical beasts on other planets?

The document to emerge from the committee, called the *Cyclops Report*, presents a 250-page elaboration on its own introductory statement: "We now have the technological capability of mounting a search for extraterrestrial intelligent life." It describes potential signs of such civilizations and a potential detection system—called the Cyclops, specced down to a capacitor—that could pick up a message. The scientists suggested the Cyclops, a telescope composed of many smaller antennas that would together have an area of 100 square kilometers, be built in a modular fashion. They could erect a few antennas and see whether any extraterrestrial broadcasts came through. If not, they would build more antennas, making the telescope bigger and more sensitive to fainter broadcasts. If they still found nothing, they would expand the array out even more. But when the federal government saw the total price tag ($6–10 billion), they put their hands on their pocketbooks and ran away. Although NASA intended the *Cyclops Report* to make SETI seem doable, the team actually accomplished the opposite.

Nevertheless, they had produced a thorough and (strange as it sounds) compelling-to-read document that remains relevant 44 years later, when the computers that ring in our pockets are much more powerful than the PDP 8/S. Some call the report SETI's bible. Tarter still gives it to graduate students to teach them to sift radio signals from outer space static. She's like a zealous convert, passing along the scripture that first inspired her.

❖

Bowyer, the professor Tarter surfed with when she crashed into the coral reef and whose class she'd TAed poorly, had read the *Cyclops Report*. Although that particular big telescope would never be built, he had a smaller-scale (and more cost-effective) idea: a SETI instrument he could stick on the back of any existing radio telescope. That telescope could do its regular work, and this instrument would make a copy of the incoming radio signal—a copy that scientists could use for SETI work, leaving the original copy intact for regular science. Without losing any information, Bowyer could do SETI at the same time that another astronomer did his or her own research. Not exactly symbiotic, and not exactly parasitic. Commensal.

"Piggybacking," he called it. Bowyer began shopping this idea around. Jack Welch soon said he had a telescope Bowyer could try as a test. It was at the Hat Creek Radio Observatory, in far northern California, off a rural logging route so dangerous that Welch had learned to fly a plane just to avoid driving there.

"You can use the telescope," Welch said. "And, what the hell, I have this old surplus computer that no one wants. I'll donate it to the cause."

It was a PDP 8/S. Bowyer didn't know what to do with that dinosaur. No one did—no one except his former teaching assistant.

❖

Bowyer walked up to Tarter's desk and plopped down the *Cyclops Report*.

"Read it," he said.

She did. And she didn't stop for two days. When she finally looked up, red-eyed, from its pages, she was like a person entering a dream world. She could find the answer to the questions that had lived in the back of her mind since childhood, as scientifically as she could probe the plasma interiors of stars.

Are we alone?

This is the first time in history when we don't just have to believe or not believe, she thought. *Instead of just asking the priests and philosophers, we can try to find an answer. This is an old and important question, and I have the opportunity to change how we try to answer it.*

Like any conversion experience, this one made the world make sense. "I just knew I'd found the right place," she says, "never having thought about it before." It felt like home. And it synced her back up with people on her own planet, those she now found herself working with and those she simply passed on the street. "There was a feeling of connectedness," she says. "I was doing something that could impact people's lives profoundly in a short period of time."

Her doctoral work then dealt with brown dwarfs—the balls of gas too small to be stars and too big to be Jupiter—and was funded by taxpayer dollars. She'd always felt uneasy about that, wondering about her research's relevance to farmers in Iowa. But when she decided to become part of SETI, her mind stilled. She walked down the street happy, knowing that her work would matter to the lives of the people she passed. It could *mean* something to them. "I didn't have to worry anymore," she says.

Without the *Cyclops Report*, which appeared in her life because of outdated computers and a botched teaching assistantship, she probably wouldn't have become a SETI scientist at all. "It wasn't in my planned universe," she says. "But chance favors the prepared mind."

And whether or not it was clear to her, no doubt the universe was unfolding as it should.

Although *should* is a misleading word: The universe has no plans or predestinations. Whatever does happen is the thing that happens, and that is the thing that should happen. And what happened is that Tarter told Bowyer she would join his fold. They called their SETI project SERENDIP—a tortured acronym meaning Search for Extraterrestrial Radio Emissions from Nearby Developed Intelligent Populations.

CHAPTER 3

MAKING THE ALLEN TELESCOPE ARRAY

t is 1 A.M. on March 10, 2014, in Northern California, and Jill Tarter's face peers from the doorway of a dorm in the Hat Creek Radio Observatory residence hall.

"It's beautiful out there right now," she says, gesturing toward the window.

Tarter walks down the hallway past the observatory's library (which is filled with copies of the *Astrophysical Journal* dating back to the 1940s) and the lounge (which houses a particularly riveting issue in a burn bucket next to the fireplace). She steps out of the dormitory and onto the cold-steeped cement. The mosquitoes that normally plague the area around Hat Creek Observatory, home to the SETI

Institute's Allen Telescope Array, have all gone to bed. The air feels clean, the planet quiet and emptied of creatures. Tarter looks up.

Above hangs a clear, dark sky. The stars look like the crafts kids make at school: crisp pinpricks in the blackest construction paper, a UV lamp shining behind them.

"I can't even find the North Star," says Tarter. "I'm going to go get my glasses."

She has hardly reemerged with her spectacles when she rushes back inside for her iPhone.

"I think that's Mars," she says, pointing at a bright reddish dot. After her celestial chart app boots up, she hefts the screen toward the sky, scanning across a quadrant. "It says Mars, Saturn, and Pluto are all near each other," she continues. She returns to staring at the sky, turning her head a few degrees at a time to change perspectives.

Each star is a sphere of plasma whose inward-pulling gravity perfectly balances the outward-pushing pressure from nuclear fusion. This symmetry is called hydrostatic equilibrium, and it's what makes a star a star. Flaring and sun-spotted, they impose their radiation on the space around them, light-years from us. Maybe that space is home to planets, which maybe are home to biology, which perhaps evolved into beings smart enough to understand starlight.

This scene is almost too easy a mirroring of Tarter's childhood memories. Looking up from the deserted beaches of southern Florida, she was certain that an alien child was looking at the sun. Their gazes, she imagined, met in interstellar space like awkward strangers' on a subway train. She has this thought over and over. It is the kind of *Groundhog Day* observation that keeps her motivated to continue the search, despite the vastness of the universe, the brevity of human life spans, and governmental desire to spend money on missiles instead of science.

A thousand yards away, the Allen Telescope Array (ATA)— Tarter's dream observatory, or at least a version of it, made for and dedicated to SETI work—scans the sky in search of intelligent life long after she has returned to sleep.

❖

The idea for the ATA came from a series of workshops held from 1998 to 2000. These posh gatherings, collectively called SETI 2020, plotted the route of SETI research for the next 20 years. While scientists and their university salaries don't expect overly cushy conferences, Tarter and her SETI colleagues didn't want those usual invitees: They wanted Silicon Valley technologists—specifically, Greg Papadopoulos of Sun Microsystems, David Liddle of Interval Research Corporation, and Nathan Myhrvold of Microsoft—on board. And Silicon Valley technologists, with their dreamy entrepreneurial visions, require minibars and rooms of their own, with views.

At the conference, the attendees' main conclusion was "SETI needs its own telescope," closely followed by "And perhaps we should figure out how to build it." Though piggybacking had worked well in SETI's early decades, to gather and handle the stream of data they expected, they would need their own setup and computers hundreds of times faster than those that existed. The tech moguls, used to thinking about the big thing *after* the next big thing, suggested the SETI scientists "make a bet on technology," specifically on a concept called Moore's law. This law, which is just an observation of what happens in the real world, states that computers double in power every two years. It has held true since the first gigantic Macintosh. Even if you haven't heard of this modern-day math, you know how slow your two-year-old laptop now seems compared to shiny new ones on the shelves of Best Buy. Papadopoulos and Liddle felt sure they could count on Moore's law: The computing power doesn't exist today, but it will be there tomorrow, when you need it. And it will be cheap(ish).

It was a Silicon Valley idea, new to the scientists, and it felt radical back then, when everyone still had dial-up AOL and videos didn't go viral because they took too long to load.

With the future's unimaginably shiny computers in mind, Tarter and her colleagues made a plan for their SETI-specific observatory:

a bunch of small antennas—350 of them—that worked together. If you point multiple antennas at the same thing at the same time, you can merge their views to make a single superior image. The telescope's combined vision will be as sharp as if it came from a single telescope 3,000 feet wide, rather than from hundreds of 20-foot telescopes spread across a 3,000-foot-wide area. It's a neat trick. After all, building a bunch of identical dollhouses and sprinkling them around a mansion-sized lot is much easier than constructing the mansion itself. "We were the pioneers of building a giant telescope out of lots of little telescopes," says Dan Werthimer, who worked on SERENDIP and the ATA and has been a leader at UC Berkeley's SETI program since its earliest days.

The team estimated the ATA's cost at $25 million. The SETI Institute just needed to find the money and a place to plop the antennas. Long-established collaborations between the Radio Astronomy Lab at Berkeley and the SETI Institute, starting with Jack Welch's 1980s offer to host the SERENDIP piggybacker, made the question of location easy: the two organizations would work on the telescope together and build it at Berkeley's Hat Creek facility. The Institute would largely build it; Berkeley would largely operate it; they would split its observation time 50/50. "The original vision for the ATA was a world-class telescope," says Werthimer.

But the money question remained. Luckily, Tarter knew someone who had $25 million to spare and a soft spot for SETI. Paul Allen, co-founder of Microsoft, had donated to a project called Phoenix in the 1990s. She asked him if he would again like to support and save SETI. While the team awaited Allen's response, they began building a prototype array under system scientist Douglas Bock: a collection of practice antennas on which they could test each new part, before transplanting them into the real-world antennas. The prototype lived in a horse paddock outside an empty barn in Lafayette, California, with the control center in the former tack room.

The night before the prototype's dedication ceremony, Tarter stayed up late, working and refreshing her email. She works a lot—seemingly all the time. "She puts in more time than anybody else on the project," says Gerry Harp, who took over as the director of the Center for SETI Research when Tarter retired and has gotten many late-night emails from her. "She's just an amazingly energetic woman. I remember that she would work long days and then go home and do email and write stuff all night long, and I don't know when she slept." (I'm pretty convinced she doesn't.)

She hoped Allen would respond with a yes, so they could announce his endorsement at the next day's ceremony. But the inbox disappointed, and she eventually headed to bed. Not wanting to wake Welch, she attempted to take off her pantyhose in the dark. But instead of being stealthy, she tumbled over and shattered her elbow. Later, X-rays in hand, Tarter tried to map the many ways she'd broken her bones.

With 13 pieces of titanium and straps embedded in her arm, the newly bionic woman sat through the dedication. "I kept wanting a more heroic story to tell," she says, and sighs. "Pantyhose."

❖

A few months later, Allen's email—a yes—came. But the yes was conditional: if the SETI Institute and Berkeley made a list of milestones—proving step by step that they could develop this telescope and its new technology without screwing up—he would support them, in installments, as they reached each goal.

"'We can do that, fine,'" Tarter says now, imitating herself back then. "We didn't know what we were talking about."

As with any large-scale engineering project, things took longer and cost more than anticipated. Tarter was frustrated.

"Whenever we or anybody would report having completed something, Jill's first question always was, 'Well, when will you get the next thing done?'" says Harp. "Everybody always felt like, 'We just did this.'"

It was frustrating to the employees to feel like their accomplishments weren't appreciated along the way. "In a way, they felt like she didn't value their contributions because it was never enough. You could never do enough," says Harp.

It became such a trope that, after a while, they just laughed it off. "But of course we had to answer the question about the next thing," he adds. They also had to fix each difficulty with the existing resources before they could ask Allen for more funds to do that next thing.

Once, Allen visited the prototype telescope to check on their progress. When he turned to the side, he found a pile of horse shit, which had previously escaped everyone's notice, right behind him.

"Well," Tarter said to him, "at least we're not wasting your money on infrastructure."

❖

After Tarter and the rest of the team—which included Harp; software engineer and telescope operator Jon Richards; astronomer Seth Shostak, who took over as the institute's co-director after Tarter retired; Dave DeBoer, the project manager; Werthimer and other Berkeley astronomers; an engineering team at Minex Engineering—met the milestones they'd laid out for Allen, they pitched him once again: you supported research and development; now let's make a telescope that doesn't live in a barn. But it's going to cost more than we thought. And, it being January 2003, Allen had just lost $7 billion on Charter Communications.

He wouldn't build all 350 antennas, he said. But if the SETI Institute constructed 200 of them, at a cost of $27.5 million, he would give $13.5 million (the remainder of the initially promised money). As a cherry, he would donate a large amount of that sum up front so they could construct a 32-antenna demonstration array in Hat Creek so the observations could start and the astronomers could try to convince the National Science Foundation to start supporting the operations. Just get other people to donate the difference, he said. He would even give some speeches to potential donors the institute lined up.

"We're pretty brash as an institute," Tarter says, "and we thought we could do it." But when they went to wealthy people and talked grandly about how the ATA could help us find our place in the universe, understand where we come from, and investigate fundamental questions of the cosmos, the response often went something like, "Well, Mr. Allen's name is already on this telescope. He's a lot richer than I am. Why doesn't he finish it?"

By Halloween of 2004, UC Berkeley petitioned Allen to shift fundraising responsibilities from the SETI Institute to the university. They were fundraising professionals, or at least had access to professionals, they said—they were part of a university with an endowment. The presentation convinced Allen, who said that Berkeley would be put in charge not only of the site and fundraising but also of the remaining batches of money.

❖

Before he fired the institute, Allen invited Tarter to attend a party—a birthday celebration for Bill Gates—on a Caribbean cruise ship that he'd had driven to Alaska. He chartered private flights for all 500 of his guests. Designers constructed a four-story stone fireplace in the main entrance lounge and redecorated the shipside restaurant every night with Alaska-themed sets. Movie stars and tech billionaires lounged everywhere. But across the vast ballroom space, Tarter spotted her own species: other scientists. "I didn't know them," she says. "But you can just recognize them." They, alongside the rich and/or famous, soaked in the one-percent life that night, sipping thematic cocktails and watching icebergs calve.

Before the cruise, Allen had given them all a task: decorate a balsawood totem pole to be exchanged during an on-board potlatch. Tarter stayed up two nights in a row, carving and painting. "It was an assignment, and I always do assignments," she says. Her days of procrastinating were over.

She chipped away at a likeness of Jodie Foster, standing tall atop a Thunderbird model car, headphones on: *Contact*. It's charming. And Allen must have liked Tarter (at least enough to see her as 1/500th of a social event) and the idea of a SETI success. But without other wealthy donors stepping into a partnership, he would soon decline to support the telescope further. "He didn't want to be embarrassed," Tarter says. Then, imitating others gossiping about him, she says, "'Oh, you know Paul and his 'Little Green Man' thing.'" If other people donated the remaining millions, no one could say he alone bet on a dud or a crazy venture. Werthimer says that Allen "just lost interest in the telescope." Allen himself declined to comment for this book.

So the SETI Institute and Berkeley attempted to go it on their own. But Berkeley still couldn't attract donations, either. And, left without the ability to build hundreds of antennas, Tarter and her colleagues used the money they had to build just 32, and convinced the US Naval Observatory to add on 10 more and potentially use them to track satellites in space.

❖

Radio astronomy, the field inside which SETI sits, has always had a strangely cozy relationship with the military. The technology that allows scientists to effectively catch radio waves from space has its roots in the most terrestrial of pursuits: war. Engineers developed radar for World War II, and that required big antennas that could both blast and receive radio-wavelength photons.

But because the technology for intercepting communications from other Earthlings—their surface radio chatter, their satellites' pings—is essentially the same as intercepting the blast from a supernova, the two sectors support each other. Decommissioned military telescopes can become astronomical instruments, and the Department of Defense can buy time on science scopes (rarely, as you may imagine, does the opposite happen).

The Green Bank Telescope sits in a 100-square-mile area called the National Radio Quiet Zone, where broadcasters such as pop stations, cell phone towers, and Wi-Fi routers are limited or banned outright. That quietude makes it easier for the telescopes to seek out signals from, say, the beginning of the universe, because they're not drowned out by earthly noise (a cell phone on the moon is "brighter" in radio waves than all but two of the brightest natural sources). But the government didn't establish this restricted zone because it loves astronomy so much—the Green Bank Telescope is conveniently located just over a mountain ridge from the Navy's Sugar Grove Station, which Edward Snowden's leaked documents would later reveal as a key player in the ECHELON surveillance program, tasked with intercepting international communications.

And while scientists don't often like to own up to their military connections, the truth is, defense departments always have more resources than scientific ones. And that can be hard for scientists to turn down, even if they don't necessarily want to help surveil who knows what or whom.

❖

When the SETI Institute and UC Berkeley first began constructing that scaled-down ATA in 2004, Hat Creek bustled with young engineers. They and Hat Creek Construction put together the bits, pieces, bolts, and pedestals that would become the first telescope dedicated to finding aliens. In a gigantic tent, two people could assemble a whole antenna from parts in four days. That tent still sits onsite today, lifeless now but still filled with the kind of heavy machinery you're not supposed to operate when impaired.

At night, the team crowded into the one-story houses and dorms hidden among the observatory's trees. They stayed up late around the Olympic-sized pool table, leaning against the pearly inlays and calculating the precise angle necessary to make each shot. Spectators watched from an orange vinyl couch that might be an Eames; no one is quite sure. A media tower of old VHS tapes still stands in the lounge like a monument to social times past: *Back to the Future* (plus sequel), the *Alien* trilogy, and, of course, *Contact*, their cardboard covers now blanched and soft around the edges.

❖

In 2007, after seven years of work, the SETI Institute and UC Berkeley sent out a birth announcement: all those rural California sleepovers had yielded telescopic offspring. Allen agreed to support the array's operation for another two years to get them started. Following the dedication and with a kick-in from the air force, which is always interested in tracking and downlinking from satellites, the ATA watched the sky constantly, simultaneously doing regular astronomy (magnetic fields, black holes, star formation) and SETI (alien civilizations, unimaginable beings, cosmic evolution), and searching for satellites for the government during its off time.

And then, in 2011, outward forces began squeezing the Hat Creek Radio Observatory, and those inside couldn't muster the resources to combat it. This was the great recession, and UC Berkeley had a billion-dollar budget shortfall. To save money, the University of California system opted to shutter Hat Creek Observatory, a scientific outpost they'd operated for 60 years. "If we had built out the array to 350 antennas or some huge number, it would have been a spectacular telescope for astronomy and for SETI," says Berkeley's Werthimer. The UC system may have hung on to that resource. But it wasn't worth the expense, the way it was.

And so, suddenly, scientists and engineers who'd invested their careers in the ATA—and the previous telescopes housed at Hat Creek, like the 85-foot telescope and the BIMA array—were suddenly estranged from it. "There were a lot of people at Berkeley who had put a huge amount of their life into the telescope," says Werthimer, noting that one scientist failed to achieve tenure because the university cut itself off from Hat Creek.

The NSF, which had operated other telescopes at Hat Creek, said the array was too small for them to take over. It wasn't world-class at its scaled-down size. And so, one last time, Tarter and the Berkeley team went to Paul Allen. Maybe he would build out those missing antennas, transforming it Cinderella-style into a telescope the NSF would be proud to splash on brochures. But instead, Allen said, "What would happen if we left you alone for three or four years?"—not providing funding or support and seeing what the SETI team could accomplish on their own.

"And that's where we left it," Tarter says. "I send him updates. He hasn't written back."

In addition to dealing with Allen's refusal and the Berkeley breakup's fallout, she was also dealing with an actual breakup itself. "We had a miserable divorce," she says of the dissolution of the ATA team. "And like any miserable divorce, it was acrimonious as hell. 'It's your fault.' 'No, it's your fault.' 'I own this widget; you own that nut.'" One night, some Berkeley engineers removed a rack of computers. They still have it. Of course, they say it was theirs to begin with, there being two sides to every story, even in science.

Still, Werthimer says hard feelings don't separate the two groups. After all, this was fundamentally an institutional separation, not an individual one. Tarter and Welch still come over to Werthimer's house often—for dinner, for dancing—and Tarter attends the Berkeley SETI group's weekly meeting. Welch still has an emeritus appointment at the university's Radio Astronomy Lab.

But without Berkeley, the SETI Institute now owned a telescope they couldn't afford to operate, like impoverished yacht owners whose boats sit at the dock all day, and they did not have the federal permits necessary to operate the site, which is on land in the Shasta-Trinity National Forest. They padlocked the gate, dead-bolted the doors, and went back to the Bay Area. The array hibernated.

But as with the ends of so many marriages, a new relationship was waiting in the wings. In the years just prior, the SETI Institute had contracted with SRI International, a scientific nonprofit organization, to see if the ATA could dip into "space situational awareness"—jargon for "what's in orbit around Earth and where." SRI developed a plan to retrofit part of the telescope's computing system with super secure firewalls. Then, clients like the Air Force, always curious about the whereabouts of satellites, could use the ATA as a tracking station. SRI agreed to upgrade from "contractor" to operator, with ownership of the former Berkeley assets at the Hat Creek facilities.

If an organization decides to close a telescope, they are legally obligated the return the land to its natural state. It's a process that often costs more than just continuing to run the telescope for years and years. The University of California knew their checkbook would be more balanced if they didn't close but instead sold the Hat Creek Observatory—even for cheap—to someone else. SRI purchased the facilities for $1.

The SETI Institute kept custody of the telescopes and computers for which they had fundraised so hard, and SRI began the proceedings to obtain a use permit from the United States Forest Service. But these negotiations took a while. And time—as it is wont to do—marched onward, as the telescope remained idle.

After a few months of back-and-forth, paperwork troubles, and continuing hibernation, Tarter drove the five hours upstate to visit Hat Creek and make sure teenagers weren't spray-painting the observatory's walls or hacking through the drywall. When she arrived, she found the desert grass growing high, creeping up on the control room. Her dream had been right in front of her: a telescope just for SETI. A real shot at finding out just how alone we aren't, or are. And now it was strangled in weeds.

❖

Even a few months later, after SRI got its paperwork in order, getting the telescope cleaned up and back on its feet required cash. So Tarter and the SETI Institute's development officer, Karen Randall, concocted a resurrection plan. They would crowdfund it, asking the masses—so fond of *Battlestar Galactica* and alien apocalypse movies—to rouse the ATA.

Crowdfunding was a new idea in 2011, the days long before you could put "I want to make a potato salad" on Kickstarter and earn $60,000. Randall developed a website, called SETIStars, just for this project. And over the course of 40 days, the crowd donated $250,000.

Tarter scrolled through the thousands of comments people had left with their donations. "'I'm doing this in the name of my two-year-old daughter because this is her future,'" says one. It felt like the thronging spectators at the 25th mile of a marathon. Jodie Foster signed it, saying, "I'm a SETIStar because, just like Ellie Arroway, the ATA is 'good to go' and we need to return it to the task of searching newly discovered planetary worlds for signs of extraterrestrial intelligence. In Carl Sagan's book/movie Contact a radio signal from a distant star system ends humanity's cosmic isolation and changes our world. The Allen Telescope Array could turn science fiction into science fact, but only if it is actively searching the skies. I support the effort to bring the array out of hibernation." So did *Apollo* 8 astronaut Bill Anders: "It is absolutely irresponsible of the human race not to be searching for evidence of extraterrestrial intelligence."

This money allowed them to unlock the gates, mow the grass, and get back to work. And that's exactly what they did—with fanfare. The Kepler telescope team, largely based at Ames Research Center, who had launched a planet-finding telescope into space two years earlier, was set to release its second catalog of new worlds—2,326 planets beyond the solar system, more than 20 times as many as scientists knew of before. NASA's Bill Borucki invited Tarter to point her resurrected telescope at the most promising of those worlds, and share in the buzz of press coverage.

"Jill and the SETI program have played a very important role in the Kepler mission for a very long time," says Bill Borucki, a scientist at NASA's Ames Research Center and the head of the Kepler mission. He worked on the design for decades before anyone believed it would work—that the planets were out there and that our technology could find them. In some ways, the search for planets mirrors the search for extraterrestrial intelligence, but sped up, and with some conclusion. And the two mutually benefit each other: the more Earth-ish planets scientists find, the more likely people are to be pro-SETI. The more pro-SETI people are, the more likely they are to support planet hunts. Both sides know that.

"It was always considered that as soon as we obtained data for planets in the habitable zone," where temperatures stay warm enough for water to be liquid, says Borucki, "that the SETI telescope would obtain data to see if there were any radio signals that would indicate intelligent beings."

SETI seemed to have an actual future, and two promising new marriages.

While the Kepler SETI search, and every search before and after, has shown nothing, Borucki emphasizes—as nearly all scientists do—that while "50 years of SETI" seems long to humans, we've only just gotten started. It's not time to give up yet.

"Wait a thousand years or so," he says, searching for signals the whole time. "Then rethink it."

❖

The ATA, in 2017, still has only those 42 dishes, which are sprinkled across Hat Creek Valley as if an angry giant turned a box of them upside down. This spread, which looks random, is carefully calculated to give the telescope the sharpest view. The antennas straddle Forest Service land and a ranch owned by the Bidwell family. A line of trees should separate the array from the ranch, so Debbie Bidwell doesn't have to look into the 42 antenna eyes all day. But the functional landscaping has been erected and subsequently eaten—twice. The predator remains a mystery (the easy joke would be to blame aliens). But regardless of the herbivore's species, the Bidwells will have to endure the view for now.

The antennas do stand in jarring technological contrast to the remote land around them. They're an hour and a half from Redding, the closest city, which is not much of a city-city but a place known for rodeos and trail-running competitions. In the town of Hat Creek, a café decorated with landscapes painted onto sawblades sells double-venti-sized milkshakes. A few miles down the road, a gas station displays at least 75 pictures of people holding it-was-this-big trout and sells a wine called Redneck Red.

The region is about five hours northeast of the Bay Area—a long, hot trek up Interstate 5, the mercury often rising above 100°F, with only yellow grass to break up the view of olive orchards. But after you veer off the highway, you enter volcano country. In the distance loom snow-capped destroyers of worlds, some still capable of blowing their tops. Mounts Shasta and Lassen, which erupted just 100 years ago, lord over the scene most visibly. The surrounding landscape ascends to more than 4,000 feet in elevation as you drive past pine tree after pine tree (punctuated by burnt pine after burnt pine) toward Hat Creek Valley. Brush rises from the fine dirt like it just landed there, while at the base of the nearest ridge, lava rubble actually *did* just land there. It cooled and formed a pile of pocked rock that stops just short of the Bidwell ranch. It would look like the remains of an abandoned construction site, if anyone built structures out of igneous rock. Off the ridge above the rubble, hang gliders sometimes jump and cruise, UFOs until the air sets them down in the valley.

The March 2014 morning after the stargazing adventure, Tarter emerges from her dorm room at 7:30 dressed for the day, bronze turtle earrings in place as always. She pours a cup of decaf coffee and stands sipping at the window in the communal kitchen, which is filled with mismatched mugs and a collection of cooking sauces left behind by engineers. Nothing is visible outside the window except the roundabout driveway. In it sits the Camry Tarter is renting until her own Saab is fixed. She rinses the mug, sets it next to the sink, and ventures out into the UV light.

She drives the quarter mile from the dorm to the main control building, winding along a one-and-a-half-lane road that doesn't afford a view until you're right in front of the telescope. The building is low, T-shaped.

"Oh, Susie planted irises," Tarter says as she walks inside, referring to Susie Jorgensen, the site manager.

I ask if Susie planted them herself.

"Yes," Tarter says. "There is no one else."

Although the site once housed a dozen staff, whose dogs kept the place lively, the observatory now has no permanent astronomers. We are the only ones here, human or canine. She walks past the window-walled computer server room and into the lobby, where sample tourist T-shirts and mugs rest beneath a turned-off TV. In a more bustling time, tourists received guided tours. Now, they push the Play button for the introductory DVD themselves. Above the scene hangs a sign showing the array of antennas: 30 DOWN, 320 TO GO, it proclaims.

Over the "30," someone has stuck a Post-it note that says "42."

Tarter turns on the DVD, a slide show and B-roll combination, produced and narrated by an intern. It's obsessive in its detail, describing the site's history like Wikipedia would. They would like to redo the video, she says, to produce something slick and flashy. But such a production sits on the back burner, way behind those 308 other things that need to be done. The credits roll, and she switches the TV off and heads down a narrow hallway, wallpapered with the kind of laminated 3 × 4-foot scientific posters astronomers make to hang for show-and-tell at conferences. They're text-heavy, with weighty titles like "Real-time Imaging" and "Launching the Galactic Center Transient Survey."

Past them is the control room, an open space with a mostly empty L-shaped desk. A whiteboard is filled with equations, their bright green and blue variables slanting slightly upward as you look from left to right. Just outside lies a doormat decorated with an iconic *Close Encounters*–style alien. WELCOME ALL SPECIES, it proclaims. Out from the doormat, someone has painted tiny green footprints on the sidewalk. They point the way from one interpretive kiosk to the next, so that the 2,000 annual visitors can guide themselves on tours of the site.

Tarter steps out onto the footprints, walking to the antennas, which shine like enamel in the sun. A pictorial sign warns of rattle-snakes. Tarter takes a magic-seeming baton from her jacket pocket and waves it in front of one antenna's support pole.

"Magnetic wand," she says.

It disables the dish so it will hold and not suddenly turn toward a star on the other side of the sky, smacking anyone in the head. The designers were considerate of cow crania, too: the dishes never aim lower than 18 degrees, leaving just enough space between the dish and the support pole for a bovine face to fit.

Each dish measures 20 feet across. It acts like a mirror for radio waves, reflecting them in the same way that your bathroom mirror reflects visible light. When radio waves stream down from space and hit the dish, it bounces them toward a second, smaller "mirror." This one bounces them again, to a detecting system. Here, the radio waves begin to turn into something humans can sense and understand, much like a car's audio system turns radio broadcasts into ballads. The smaller second dish protrudes from the bottom lip of the main one, stuck out on support poles like a chin. A metal shroud covers the lower space between the two, and a Sunbrella covering straps over the top.

Sailors are familiar with Sunbrella because their sail bags, chairs, and booze-cache covers are made of it. It keeps water out but lets radio waves through, much like a skylight does visible light. Tarter points to a thick seam of Sunbrella threaded through a slit in a pipe, which keeps the fabric closed and flush against the main dish, a tech-nique learned from her sewing years. This is a keder welt, something sailors use to raise sails up a mast.

She stands on tiptoe and struggles to undo a latch on the metal shroud, trying to get into the antennas' insides. It's a rubber hook and eye designed to hold truck hoods closed. This sort of

MacGyvering—which seems like it resulted from tipsy late-night Amazon shopping—appears all over the telescope. The array is a maker scope in that way, similar to the do-it-yourself creations popular in tech-nerd circles: homemade Geiger counters, coffeepots that brew when you flip a light switch in your room.

"Come on," she says, still trying to open the trapdoor. "Come. On."

But the latch remains just out of reach. Eventually, she retreats to the brush, where she roots around until she finds a stick. Holding it aloft like a trophy, she extends it toward the antenna and coaxes the rubber loose. The trapdoor swings open so fast it seems like a jack-in-the-box should pop out.

"Aha," she says.

It is a Stone Age triumph, a reaffirmation that humans can augment their abilities with tools, like sticks or telescopes. She pokes her head inside the antenna and looks around as if orienting herself. Next to her, a pyramid-shaped piece of metal protrudes from the main dish, peak pointing outward. Tapered rods—like those TV antennas people used to place on their roofs—adhere to each of its four sides. At the pointy end, they're half an inch long, but they get bigger as they progress down from the pyramid's peak. It looks like a death ray.

But it's not a death ray. Instead of blasting rays, it detects them. The ATA picks up signals 10 to 100 times higher than those your radio receives, from about 1 to 10 gigahertz (imagine a station called WMFE 1000.3). Each different-sized rod picks up, essentially, a different cosmic "radio station." Your FM car radio is sensitive to radio waves with frequencies of 87.5 megahertz to 108.0 megahertz. "One of the problems in SETI is you don't know what frequency ET might be broadcasting on," explains Werthimer, "so you want to look at as many channels as you can."

But radio waves, like car headlights, dim the farther they travel into the distance. And signals from space have generally journeyed trillions of miles before they get to us. They arrive weak, in need of assistance. SETI scientists don't want to miss the alien signals simply because the telescope wasn't sensitive enough, the broadcast too bedraggled, the civilization too far away. So a circuit that Tarter's husband, Jack Welch, designed, amplifies the signal. Welch's amplifier circuit makes message-finding possible. The couple have been working on the ATA design together at their dining room table for years. Welch's notebooks and papers are always piled up, pushed aside for dinners.

"Do you think it would change people's day-to-day lives if we found a message from extraterrestrials tomorrow?" Tarter asks. "I wonder," she continues, nodding to herself and walking toward the green-alien doormat.

❖

Back inside the control building, someone has taped a piece of paper with a printed gremlin on the server room door. Its impish gaze follows as Tarter opens the door. Inside, racks of computers hum and heat the air. A/C units do their best to keep the temperature at a cool equilibrium. The room has more optical fibers and wires—plugged from this into that, and from that into something else—than a Radio Shack warehouse. They are thick, thin, bundled, loose, curving, dangling, sagging, taut, blue, orange, purple, green, yellow, red. They have invaded. Tarter slips around the Medusa's head of wiring.

At the back, the processing systems have labels: PRELUDE, SonATA, and FOXTROT. She shakes her head at the latter one and mutters, "IT guys," under her breath. In the early days of SETI, those guys custom-built every bit of this hardware and software. Bespoke may be cool for fine Italian suits, but scientists would prefer it if they didn't have to make their own equipment. Today, consumer technology is finally powerful enough that SETI can steal it (just as the SETI 2020 technologists predicted). When Dell, through vice president Forrest Norrod, and Google donated ultra-fast servers for the ATA, the SETI Institute finally got to throw away the ones they'd spliced together.

But certain parts are still custom tailored, like a piece of hardware labeled BEAMFORMER, which allows the telescope to zoom in on specific exoplanet systems and search them for intelligent inhabitants in real time. (Extraterrestrial broadcasts, like our own, might turn on and off. If the ATA picked up a "Hello, Earth" broadcast on July 10, 2014, but no one noticed for a week, the interstellar show may have ended already.) The SonATA software figures out which distant worlds the ATA can see (its view spans, at any given time, a patch of sky as wide as five of our full moons), and homes in on three of them at once, looking for broadcasts that occur at a single frequency, like our radio stations.

In this fluorescent-lit room filled with fiber and fans, a computer program searches autonomously for evidence of extraterrestrial life. Normally, this search runs itself, and this room is empty. But today, the room's blinking LEDs reflect off Tarter's jewelry, bouncing like a beacon from the turtle pinned to her earlobe.

❖

Back in the control room, Tarter flips open her MacBook Pro, which somehow stores more files than most people create in their entire lives. She navigates to a website called SETIQuest, which shows the ATA's progress: how many planets it's searched for signs of smart life. On one page, a GIF of Inspector Clouseau, tapping his fingers in impatience, pops up. When observations start up again, the detective disappears and a buxom woman takes his place, peering back and forth with binoculars. Jon Richards, who wrote the software that runs the telescope and also, usually, runs the telescope, created the page and chose the animations. "Each of us has our job to do," says Richards. "Mine is to keep the signal search going."

That's kind of the most important SETI job at the SETI Institute. And so he gets to design the webpage and make animated GIF jokes if he wants to.

Tarter sighs. "I've been thinking we should age the woman," she says. And, indeed, Richards later upgraded the website, giving it new interactive features that allow viewers to follow along and access real data from the day's observations. Buxom women (of any age) and French inspectors have disappeared.

Astronomers have confirmed around 3,500 planets as of January 2017, and they have thousands more candidates awaiting confirmation. When the ATA points at these exoplanets, collecting whatever radio waves may be coming from their direction, the software scans 9 billion different radio channels in search of an alien signal. Each of these channels is a tiny slice—a single station—of the ATA's "radio dial." Scrolling through all the planets and all the frequencies will take at least five total years of dedicated time.

The list on the site displays worlds upon worlds, like some kind of yearbook full of strangely named students: Gliese 581, Kepler 69, 51 Pegasi. Which is the most likely to succeed?

❖

Each target gets 90 seconds in the spotlight. If SonATA doesn't see anything that looks suspicious, it moves on to the next candidates. But if something catches its attention, it halts, about-faces, and assesses the likelihood the rogue radio waves come from ET. Has this particular signal ever shown up before? Is it coming from just one place in the sky? If so, SonATA checks the planet again—and then three more times—before any humans find out. If any signals survive those five trials, SETI scientists' cell phones ring and beep with text messages.

The responsible researchers log in from offices or bedrooms to continue to follow up the signal. If it persists, they drive up to Hat Creek. They then check the signal themselves and must withhold judgment (and celebratory Champagne, which sits always on ice) until they've run out of tests or the exoplanet begins to slink below the horizon. If the signal still looks promising, the scientists phone another observatory, begging the director for "discretionary" time—emergency follow-up observations that don't have to go through the usual lengthy process of approval. Do they see that same signal? Does it look like a message from intelligent beings?

They have gotten that text message exactly once. And the signal, as usual, turned out to be just evidence of us.

❖

That evening in March 2014, Tarter drives to the nearby Burney Falls, one of her favorite local views. She walks down the Conservation Corps' stone stairs. Just like this is a family vacation, she stops at each interpretive poster, learning about the volcanic dynamics that shaped this region. At the falls, which dump 100 million gallons of water over their edge every day, she observes in silence. It is beautiful: a cascade pouring over a wall of basalt and through the many holes in the rock. Tarter points all the way across the gorge to a tree that towers over the whole scene.

"There's a nest up there," she says. She proceeds to wonder what type it is and whether the raptor ominously riding air currents lives there. Such speculation normally is idle: wonder, move on, and forget. But when the ranger's office opens the next morning, Tarter calls to inquire about this bird's species: a bald eagle.

She wants to understand birds' nests and lava formations just as she does fast Fourier transforms and Nyquist sampling. On some rocky Earth-twin world a few hundred light-years away, perhaps a bizarro Burney Falls exists, formed by that planet's own fiery interior and bored through by the relentless flow of water. Perhaps she is thinking about that.

Perhaps she is not. Maybe the world around her is satisfying on its own, right now.

That night, a few thousand stars again are visible. It's impressive, but at least a hundred billion more are out there, in our galaxy alone. Astronomers now estimate the Milky Way is home to at least 100 billion planets, too. And as *Contact* aphorizes, if there's nobody else out there, "that seems like an awful waste of space." But it's precisely because space is so big that the search for extraterrestrial technology is so hard and uncertain. Tarter may hope, every time she pulls out her iPhone to consult a celestial map, that an alert from SonATA will pop up. She may wish that the software would summon her colleagues, gathering them back to the Hat Creek control room to white-knuckle through the confirmation protocol. But she knows, and they all know, that with the universe's nearly limitless possibilities, the call could come in 10 years, tomorrow, next month, or never.

Perhaps with the telescope's new receiver upgrade, slated for installation in late 2015 and being constructed in Antioch, California, the ATA will find aliens whose home lies much farther away.

CHAPTER 4

THE FUTURE OF THE ALIEN-HUNTING TELESCOPE

Antioch is the home of the milk carton. In the 1950s, visionaries at Fibreboard Research Company figured out how to apply wax coatings to cardboard containers, saving schoolchildren everywhere from dairy products that taste like paper. Now, however, Antioch is home to another innovator: Minex Engineering Corporation, which creates precision parts for the Allen Telescope Array.

Minex lives next to a property management company, at the end of an office park cul-de-sac called Apollo Court. The yellowish-gray industrial siding and single unadorned glass door make you think this must be a different Minex. Not the Minex where people develop the technology to find ET, but maybe the Minex that does your taxes or mails packages for you.

But it is at *this* Minex that engineers are currently working on the upgrade that they hope will make the ATA's vision twice as good. Instead of building more antennas, the team are making the existing antenna better. They designed a new "feed"—the pyramidal part that turns the radio signal into data humans can understand.

Inside this pyramid, the air is a hyper-cooled vacuum. The colder that vacuum is, the less the atoms in the electronics move, and the stiller they stay, the less they interfere with scientific signals. The upgraded feed will have a similar pyramid but finer electronics and will live *within* a larger vacuum chamber. In other words, everything will be colder and quieter, and the telescope will be able to pick up weaker signals from space because it won't be making as much of its own noise.

If the team can successfully grow and transplant this new organ into the 42 antennas, each antenna will become twice as good at its job. But in July 2014, when the Minex engineers meet with some SRI International staff to update them on the progress, it doesn't look good. The upgrade is over budget and off deadline, and the warehouse is 95 Central Valley degrees. The whole project might be impossible.

Inside Minex, Welch awaits the beginning of the meeting. He sits asleep at the reception table, which is just a card table in the middle of a square, white-tiled room. A notebook of hand-scrawled calculations lays open on the table next to him. The inner door swings open as two men, Minex engineers, walk into the lobby. The brief gap between door and wall gives a glimpse of CNC millers, anechoic chambers, wave solderers, and other jargony devices that look like someone might actually create telescopes here. The men wear jeans and name-emblazoned work shirts. Matt, one says. Brad, claims the other.

"Jill here?" they ask.

Welch does not wake up at the mention of his wife's name. His head still rests against the chair; his hand still sits limp against his logbook.

"I don't see any SRI guys," Matt says.

He retreats to the back room. I follow. Against the far wall, a large shipment of molded plastic languishes. The brown, columnar molds look unambiguously phallic, all lined up and ready for action. This high-tech plastic, usually used for fighter-jet components, was supposed to sheath the antennas' new insides. This foray into the world of exotic plastics came about when the price of fused quartz glass, the material initially proposed for the vacuum jars, jumped radically after funding was obtained. The molds let radio waves pass through but seal off air from the outside world, maintaining the ultra-cold vacuum inside. Matt picks one up and explains that the material outgasses, meaning that tiny molecules leak from its sides. Matt and the others tried painting a seal on, but that didn't work, either.

Above the shop hangs a thematically relevant fighter-jet calendar and a less on-topic fast-car calendar. A wheel-bottomed organizer houses drawers labeled !!! DON'T TAKE IT; PARTS, NOT CLEAN; and LOCTITE KRYTOX, which sounds like an alien language but is actually a type of lubricant.

Just as Matt puts the phallic plastic back where it belongs, Tarter walks in, husband trailing behind her. Sunglasses still shade her eyes, and red toenail polish peeks through her sandals. She brandishes an email she sent earlier that morning. Subject line, straightforward as a *For Dummies* manual: "Charts that help me think about what we have achieved." The body of the message has the structure of a scientific paper: a list of six charts, an observations section, questions for discussion, and conclusions. The charts show how the electronics inside the updated feed should perform compared to the old model, as well as how they actually *do* perform. The new instruments will be sensitive to a whole range of high frequencies that the old instruments could not detect at all—and more sensitive to even the lower frequencies the old ones could detect. The upgrade will give them access to parts of the spectrum—and so potentially to alien broadcasts—they could not have seen before, and will let them pick up fainter messages of all stripes.

"Everybody here?" she asks.

The SRI guys meander in shortly after. They wear suits and have clean-lined haircuts.

"We're all here," they say.

❖

The gist of the engineers' presentation is this: the upgrade is all wrong. There's the outgassing, sure. But atoms also leak into the vacuum. And the system slowly warms up over time. They have no idea why.

"Dark matter," Tarter mumbles. It's a sort of joke. It gets a laugh from a young guy who appeared out of nowhere and hovers silently in the back with his arms crossed over his half-tucked-in silk shirt.

If they can't fix the fighter jet plastic, Matt says, they'll have to use glass covers. These covers are pretty and more reliable, but expensive—just like a glass bottle merits a greater return from the recycling bureau than a plastic one. And on top of the extra money for the material itself, there's the difficulty of making the pyramid's tip super thin—just 40/1000 of an inch. No company thinks the delicacy required to make a glass pyramid that thin is worth it, not even for a lot of money.

"How thin will the company agree to make it?" Tarter asks.

"Sixty thousandths," Matt says.

Tarter turns to Welch. "How much signal do we lose if we accept that thickness?"

Welch consults his notebook. "Five percent," he says.

"On top of what you lose already by having forty thousandths?"

"With forty, you lose three percent," he says.

"So, we'd lose an extra two percent," Tarter counters.

"Well, you'd lose five percent total."

Tarter sighs and says, "Okay."

The engineers shift their weight and look around at each other. Their colleagues sometimes have separate meetings with Tarter and

Welch, so they can relay information between the spouses, rather than having the spouses spat.

Later, Tarter and Welch will have an argument about which electrical component company makes some electronic component.

"Hittite," Tarter will say.

"Right," Welch will continue, "those Hitachi things."

Tarter worries that Welch's memory is slipping. She sometimes lashes out in frustration, picking little fights about the inconsistencies, or shaking her head and correcting him. In his potential slippage, 10 years biologically ahead of her own, she sees her future—and fights against it. She tries trying to force him to remember "Hitachi," as if by doing so she can ensure she will remember it when she is 80.

"Never get old," she says to me, often.

But the engineers have already left the familiar marital bickering behind and moved into a discussion of electronics: the feed's low-noise amplifiers. These boost the signal, essentially turning up its volume. Matt lays pages of price quotes out on the lab table. The comparisons come from a place called the Low Noise Factory, suggesting an underworld market invisible to the outside world. Trouble, though, is also afoot with the amplifiers. Neither of the two options performs as well as they should when they attempt to amplify radio waves at higher frequencies.

Tarter places her hands on the table in front of her, her head hanging down. After a few seconds, she sighs and looks up. "We have to have feeds at some point," she says. She gazes around at the crowd, who are silent.

"Really," she continues. "We have to have feeds. At some point."

But the thing is, they don't. There's no guarantee the SETI Institute can make the feeds work as planned, just as there was no guarantee they could build 350 antennas, which they haven't so far. Given the time delays, the budget overruns, and the failure of most feed components to behave like they're supposed to, one has to wonder if the ATA and its upgrade will turn into SETI's version of Boston's Big Dig. How far should they go, and how much money should they spend to make this particular telescope happen? It's not like it would be first failed or canceled scientific instrument—the Superconducting Super Collider, the Tevatron. You never believe *your* scientific instrument will be the one to fail, but somebody's has to be.

But, so far, no one at the institute has yet thrown their hands up or the towel in on the ATA upgrade. And the team decides, months later, to go with the glass covers. They accept the slightly muted radio waves. And a few months later, they solve the temperature and amplifier problems. By the summer of 2016, they have finished designing and have started installing new feeds—10 in total. Things start to look up. Although they then ran out of funds to complete the installation, an influx in September will bring the total to 20—nearly halfway there. Tarter and Welch, in August, sold their share in the small plane they co-owned with two young partners and donated the money to the telescope's improvement.

❖

The ATA's upgrade was mainly funded by a man named Franklin Antonio. On November 14, 2012, the SETI Institute announced that Antonio, the co-founder and chief scientist of communications giant Qualcomm, had given $3.5 million to the cause. An engineer himself, he likes to be briefed on (minute, monotonous) details in paper form first, so that he can dial in to any teleconference fully informed and with intelligent questions on how his money is being spent.

Antonio and six other engineers founded Qualcomm in 1985, becoming one of the early Silicon Valley legends. The company began by helping long-haul truckers communicate with each other by satellite, and it evolved into one that provides your smartphone with that nice LTE next to its 4G. They specialize in wireless, Internet, semiconductors, and everything else Silicon Valley stood for before it stood for apps.

In that way, it makes sense that Antonio would fund the venture. Maybe looking for communications from aliens is the next logical step in world telecomm domination, now that cell coverage and mobile Internet are on lockdown. On the other hand, his donation pattern doesn't quite add up, and his ethos doesn't quite fit with Tarter's. He has donated thousands to the Republican National Committee. China recently hit them with an antitrust suit. Antonio hardly lives in the same left-leaning, one-Earth world Tarter inhabits, or the one most SETI scientists—who are on average more liberal and dreamy than the average person—occupy. But because he gets the goal of finding ET, this may be the definition of a time when beggars don't permit themselves to be choosers.

❖

Tarter wants to use Antonio's feed as an olive branch, extending it toward the regular radio astronomical community. The radio astronomy community is, on average, less antagonistic toward SETI than other astronomers are. SETI scientists and traditional radio astronomers both primarily use radio telescopes—from those in Green Bank to Australia to Hat Creek—to detect radio waves from space, so the data collection, instruments, and analysis are often similar. But radio astronomers' radio waves come from inanimate objects like supermassive black holes and supernovae, while SETI scientists hope for signs of animated, conscious sources. But they can meet against the walls of parties to discuss the latest additions to the Low Noise Factory. They generally all started out in traditional radio astronomy and had the same education, before the SETI scientists veered off in their own directions. Tarter calls radio astronomer "her people," and these people did award her their highest prize in 2014—the Jansky Lectureship, given to one radio astronomer per year. "Jill is being honored for her role in pioneering methods for searching for extraterrestrial intelligence using radio techniques, as well as her leadership in the emerging field of astrobiology," Tony

Beasley, the director of the National Radio Astronomy Observatory, said at the time.

However, while SETI scientists use radio astronomers' telescopes, radio astronomers rarely use SETI's telescope. More than 300 antennas short of where they planned for it to be, it's just not as good as other bigger, better facilities at its current specs. Those larger facilities usually have federal funding, because governmental agencies see them as benefiting the scientific community as a whole with reliable results released regularly, whereas historically the government has been reluctant or, more often, outright hostile toward funding something as fringe as SETI.

Tarter would like to convince astronomers, though, that they can do good traditional science with the upgraded ATA, while the SETI Institute does SETI in the background. The bait? A newly discovered mysterious phenomenon called a fast radio burst, or FRB, which Tarter calls a "furby." SETI scientists and regular radio astronomers are both obsessed.

❖

In 2007, Duncan Lorimer and his partner, Maura McLaughlin, sat at home sifting through old data from the Parkes telescope in Australia. In that old data, Lorimer found something strange: a huge, single, singular pulse of radio waves. Lorimer told McLaughlin to come look at this crazy thing, which just appeared as a big black splotch on the graph, tapered at both ends like two teardrops stuck together. The burst had lasted just 5 milliseconds, but in that tiny amount of time, it had released more energy than our sun does in a month. It had traveled, they calculated, 3 billion light-years before it arrived at Earth. During the initial excited speculation, other astronomers threw together papers explaining how something so energetic could happen so fast so far away. The theories were all big and bangy: a flare from an ancient black hole, cosmic strings, unknown dynamics inside supernovas, supermassive stars smashing into each other. It became known as the Lorimer burst, because only Lorimer ever saw one. The situation was like living in a kids' book, where the children know Narnia (or fairies or elves) exist, but every time they try to show their parents, the fairy disappears. Speculation turned to doubt, and "What is it?" became "Is it really?"

Four years after the initial discovery, an Australian research group led by Sarah Burke-Spolaor suggested that Lorimer had been fooled: his bursts actually came from Earth. Lorimer didn't believe it was true, but many others did. Scientists later discovered that the bursts Burke-Spolaor saw came from microwaves. When astronomers at the telescope site opened the microwave door a little early, without pushing Stop first, a flash of radiation flew from the open door before the mechanism shut off.

Then, in 2014, Burke-Spolaor found a *real* burst herself (though the name *Lorimer-Burke-Spolaor burst* didn't catch on). Soon, another team discovered four more. They changed the phenomenon's name to fast radio bursts. One of these even burst into the sights of the Arecibo radio telescope in Puerto Rico, meaning that FRBs weren't just a fluke. And in January 2015, scientists in Australia announced they had watched an FRB happen in real time using the Parkes radio telescope. They had pointed the telescope at a place where nothing was bursting, and then something had burst—3 milliseconds, a day's worth of solar output. Since then, astronomers have found a total of 18 sources of these weird bursts, including one that repeats itself. In January 2017, they were able to localize that repeater, showing that it came from a tiny galaxy 3 billion light-years away.

But even today, and with that one known galaxy source, no one knows what object inside a dwarf galaxy might shoot these FRBs into space, and whether the maker of the 17 others is the same.

At the beginning of the FRB saga, Tarter found the bursts suspicious—possibly SETI-style signals. "Engineered signal," she says.

She means the bursts seem designed—like a beacon, or a cosmic lighthouse. *NOTICE ME!* they scream. *I look almost-but-not-quite natural!* That's a quality Tarter and others have long said extraterrestrials' messages might have. If aliens make signals that are fraternal twins of natural ones, traditional astronomical tools will detect the signal just in the course of doing their everyday jobs—and then hopefully terrestrial astronomers will notice that something looks a little suspicious.

But once more popped up, from parts of the universe very distant from each other, Tarter thought aliens an unlikely source—after all, why would civilizations across the cosmos send the same type of missive? The astronomical community, some of whom once considered semi-seriously the idea that the bursts came from a *who* and not a *what*, agrees with her new opinion. "I think that it's unlikely that these are from aliens," says astronomer Maura McLaughlin of West Virginia University. "It would take a *lot* of energy for aliens to make a broadband signal. They'd need to harness the energy from many, many suns. It'd be a lot easier for them to make something bright and narrowband." It would be easier, in other words, for them to make the kind of FM station–style broadcast that Tarter has focused on finding since the earliest SERENDIP days of SETI.

Still, radio astronomers looking to find more FRBs can use SETI-optimized telescopes like the ATA—which have wide fields of view and can splice signals into tiny chunks of time to look for on-off flashes, whether they be from Alpha Centaurians or unknown objects in distant galaxies. And they can do so over a wide range of frequencies.

Because of those capabilities, and whether or not the ATA ever becomes the world-class telescope its creators imagined it would grow into, it has been an important R&D project and a test bed for the next generation of observatories. "The concept has already come to fruition," says Werthimer. Its large number of small dishes construction is the way of the future. The world's largest telescope, which the international Square Kilometre Array collaboration is building, will have thousands of antennas spread across thousands of miles in South Africa and Australia. Like the ATA, it will swallow huge chunks of the radio band and split them into tiny frequency pieces, at tiny time intervals. It will see signals that burst, as well as ones that stay lit up, and will know how wide or compressed they are in frequency. The Square Kilometre Array and its predecessor and prototype, MeerKAT, both owe a great deal their instrumental development to the ATA—the first telescope to collate and process such fine time and frequency data from so many antennas.

And so, while Tarter is spending her retirement years (and assets) trying to make sure the ATA survives, Werthimer contends that even if it dies off, its children will thrive.

CHAPTER 5

A QUESTION FOR OUR TIME

B ut before SETI had its own struggling telescope at Hat Creek, the scientists let their instruments piggyback on other Hat Creek telescopes, as they did in Tarter's first SETI project: SERENDIP. Just a few months after Professor Bowyer gave Tarter the *Cyclops Report* in 1972, he had lined up his list of SER-ENDIP essentials—someone to talk to his ancient computer (Tarter), a computer to deal with the data, and a telescope. He needed only one more thing: money. And a man named John Billingham, soon to be the head of NASA's Extraterrestrial Research Division and former head of Project Cyclops—had plenty.

Billingham had been interested in searching for life in space since his earliest days in the space agency. He'd heard talk in the hallways about the strange-named field of exobiology—*exo* meaning not of this Earth. Scientists now call the study astrobiology, and today it mostly means the search for non-technological signatures of life on other planets: biosignatures. For example, if astronomers see oxygen (O_2) and methane (CH_4) together, that's suspicious: methane could mean farting fauna or volcanoes; oxygen can come from photosynthesis or from when ultraviolet light strips carbon dioxide into its constituent parts. Astronomers haven't found a non-biological scenario that puts *both* methane and oxygen into an atmosphere.

But Billingham's awakening interest in space biology wasn't about molecules and microbes. He was interested in the smart stuff: He wanted to learn how to find intelligent extraterrestrial civilizations that could communicate with Earth. That's why he and Barney Oliver, the Order of the Dolphin supporter who had flown in to Green Bank to spy on Drake's experiments, had led Project Cyclops. After that initial work and the federal decision not to build the telescope, Billingham had gone on to convene six NASA workshops on interstellar communication, one on cultural evolution, and two on the detection of planets beyond the solar system. With his interest in the topic of aliens and his disappointment at the lack of governmental investment in what he took to be a fundamental scientific question, Billingham seemed to Bowyer a likely candidate to support the low-budget, low-risk SERENDIP project.

So Bowyer decided to pitch him, in style. And so it was that Tarter ended up on a regional airport runway at O89 in Fall River Mills, California, during the closing days of 1972. Jack Welch, who was also a pilot in addition to being the head of the Radio Astronomy Lab, had rented a Cherokee Six airplane for the big day. The three of them were flying Billingham up to the Hat Creek Observatory to show him their equipment. *Whisk* was the verb Bowyer used. Billingham could stand under the 85-foot telescope to which they would attach their alien-hunting instruments. They would bedazzle him with their anachronistic computer.

The team whisked Billingham from San Jose to Hat Creek. They whisked him into the town's finest diner. They whisked him around the observatory site. They gave him the chance to put his money where his white papers were. And then, the day nearly over, Bowyer whisked him to the local liquor store to get drinks for the flight home. Mixology, surely, was the way to a NASA manager's heart.

Bowyer plopped himself and the booze at the back of the plane and sent Tarter to sit up front with Welch, upsetting the proper weight and balance distribution for the small plane. Welch looked dapper in a cap and sunglasses. Tall, kind, gentle, smart, and funny—the usual good qualities, the ones people always list when someone says, "Why do you like that person?" She buckled her seatbelt as Bowyer poured, soon taking the drink he handed her.

Welch started the plane and began taxiing down the runway, lifting the nose off the tiny tarmac and pointing it above the mountains that encircled the valley. Tarter felt that extra little lift right after takeoff, the kind that shows up in your stomach like an emotion. The telescope dish receded beneath them. The whole observatory site started to look like dollhouse astronomy, like they'd all just been playing at being scientists. But it was far from playtime for Tarter. SERENDIP was the most serious project Tarter had ever been part of since she completed her PhD. And maybe, she thought, it might actually come together.

Then a beeping started sounding from the control panel. Tarter knew what the noise meant: it was a stall warning. She watched Welch's face, in the same way that commercial passengers survey their peers to see whether they should panic at big turbulence. But Welch didn't even blink, so Tarter pretended everything was normal, just like a good passenger, while Welch lowered the plane's nose until the stall warning silenced. And slowly, slowly, he spiraled and spiraled, climbing them to cruise altitude.

"Who needs a refill?" Bowyer asked, as they cruised back toward the Bay. He hadn't even noticed the alarm.

Although Billingham didn't bite that day, he would support SETI in the future, from starting a systematic SETI program at NASA to serving on the SETI Institute's board, and would become one of the most important people in Tarter's life. Welch would, too.

And SERENDIP I, which did come to fruition without Billingham and got funding in 1976, after a few years of tiny grants and volunteered hours, has continued up to a sixth iteration, a commensal instrument that sits inside the Arecibo radio telescope to this day. It's much more sophisticated than the original instrument, says Werthimer.

"SERENDIP listened to one hundred channels at once," he says. "At the time, we thought, 'One hundred channels—that's really cool!' Now, of course we look at one hundred billion channels."

Werthimer recalls that, during the SERENDIP days, a female engineer was a strange being to have on an engineering project. "People told Jill she wasn't going to make it because she was a woman," he says, referring to engineering and science colleagues.

❖

Tarter still goes to Hat Creek, sometimes still in planes with Welch, to whom she's been married for 37 years. They frequented the restaurant closest to the observatory—the Bar K, which recently closed—which they loved for its supersized milkshakes and its post-and-beam homeyness. They were friends with the owners—Jack and Donna Garner—who worked as much as Tarter and Welch even though "they're really old," says Tarter. But every summer, Welch stayed at their house in Berkeley while Tarter took student interns to the Bar K instead.

These interns are part of the National Science Foundation's Research Experience for Undergraduates program, which pairs undergraduates with established scientists for a summer research project at any of dozens of sites across the country. The "summer students," as most institutions call them, usually live in dorms at their host institution. The SETI Institute interns live in old military barracks at NASA's Ames Research Center.

They spend most of their 10 weeks working at the SETI Institute, stuck in the middle of Silicon Valley suburbia along with Ames. But for one week, they travel in Astro vans up to far northern California. Here, they get to use the ATA themselves, explore the landscape, cement their budding and doomed romances, and learn to cook for themselves en masse. And so it is that Tarter ends up at a Safeway in July 2014, paying for ten jars of pasta sauce, pounds of ground beef, and seven bags of spring-mix salad as part of seven carts of groceries. She uses the SETI Institute's credit card, which the bank then freezes because buying ten jars of pasta sauce doesn't fit with the institute's usual pattern.

Afterward, the students put the groceries away and gather in the telescope's control room. As they wait for their turn to operate the ATA, they turn to artistic pursuits. On a whiteboard, they draw nearby Mount Lassen in the process of erupting. Their dead or dying bodies, each with identifiable hair and clothing, litter the landscape below the volcano. The image's title appears to be "BOOM." The next day, they will hike up and down another inactive volcano inside Lassen National Park, visit a lava tube, and watch steam rise from acidic hydrothermal vents. They will learn about the geology of an active Earth, which is meant to give them insight about the surfaces of distant planets. On those planets, which might seem inhospitable at first glance, there could be weird life that might survive in the hot, low-pH pools there, as it does on Earth. In the 1980s, scientists started discovering these hardcore life forms—extremophiles that survive in the Mariana Trench, the deepest oceanic spot in the world, or a half-mile underwater in an Antarctic lake.

This whiteboard drawing is the students' vision of their future exploring one of these extreme earthly spots.

Tarter's own research summer intern, Lindsay, perches on the windowsill, watching a thunderstorm emerge from the nearby mountaintops. The clouds look like the fingers of an outstretched hand, reaching over still-snowy peaks, creepy.

Tarter peers out the window. "I hope that doesn't come here," she says. "We don't need any lightning."

In the parched and brushy landscape, thunderstorms aren't a nice excuse to sit on the couch reading. They mean wildfire. Most of California looks like a matchstick. The state is in a years-long drought. All along the interstates, flashing LED signs proclaim "Severe drought, help conserve water." Almond farmers are burning their trees so they don't suck water from the ground, and San Francisco has stopped irrigating its medians. Much of the state is yellow, brown—sick-looking but somehow still beautiful.

❖

The next day, when the interns return from their five-hour hike up and down a cinder cone volcano, sweat has dried in streaks through the dust that coats their calves. They rush to the trailhead's spigot to refill their 16-ounce Dasani bottles, long ago empty. It is hot, and prior to today they have spent their whole summer in front of computers. They are not in the best shape.

A horn honks, and Tarter's arm waves from the driver's side window of her Saab, parked under a pine tree. She opted out of the hike because she was afraid of hurting herself when she has to take care of Welch, who is awaiting back surgery. She's been reading in her car all afternoon, moving her car every hour or so to follow the shady patches cast by pine trees. "Some woman with a bird came up and asked me if she could have my spot because she had a bird in the car," she says. "I said I needed it. Who brings a bird to a park?"

The students wave back to her and return to their passenger vans, ready to head to the next Earth wonder on their educational journey. The vans leave, dipping and turning on the mountain roads in front of Tarter. One hand on the wheel, she twists the car around an S-curve. She's a speeder. (Once, during a previous intern trip, a police officer pulled her over for speeding. The officer let her off because a student told him that they were all "looking for aliens," and he took that to mean illegal immigrants, a search of which he apparently approved.)

The students' next stop is another volcanic phenomenon. Molten rock long ago burned a pipe—called a lava tube—into the Earth. It looks like a gigantic worm tunneled through the ground, wiggling this way and that on a quest for food, shelter, or escape from whatever chases giant worms. The floor, uneven, plays tricks on the ankle joints, and Tarter hangs back with Lindsay, who twisted hers the week before.

"What's that?" Lindsay asks, swiveling her headlamp down to the ground. A circle of light soon illuminates a Dum Dum wrapper.

"Why do people do that?" Tarter asks. She bends down and brings it into the beam of her own lamp. "Do we have somewhere to put this?"

For the rest of the hike, Tarter and Lindsay diligently search for litter—subtle signs that questionably intelligent life was here—and place it in the cargo pocket of my hiking shorts. Glass, toothpicks, Ziploc bags.

"Look," Tarter says in the blackest middle of the cave, where the total darkness makes you hallucinate twinkling lights. She brings a scrap into the circle of her headlamp. "A roach clip," she proclaims, handing it to me to put in my pocket.

When we're halfway through, the interns play a trust game. It goes like this: Everyone turns off their head lamps, teams of three people link arms, and the person closest to the wall puts their hand there and leads their partners through the darkness. It is a game a person might have played at church camp, with a different metaphor. Tarter grabs the arm of Gabe, a curly-haired outdoorsy kid who's tan and wearing a tank top. She whispers, "Biggest and strongest. Good for balance!"

Gabe's steps pull the center of gravity ever forward. Our team travels forward quickly, confidently, like those blind albino newts that evolved optimized for places exactly like this. Aside from the feeling of the ground beneath our feet, it's almost like being out in the space, surrounded by air so black it seems thick.

In this kind of darkness, it's hard to tell if there is anyone else out there. But I constantly feel like someone is right in front of me, and I could bump into them at any time.

❖

A few weeks after this trip, lightning strikes twice near Hat Creek, starting what the Forest Service soon calls the Eiler and the Bald Fires, which burn a combined 62,000 acres. The Bald Fire scores the mountains just behind the ATA, turning the horizon into an eerie permasunset. Its flames look like the erupting volcano the students drew, bearing down on the telescope. The Eiler Fire creeps along Highway 89, taking out two houses, 12 outbuildings, and the beloved Bar K restaurant that Tarter and Welch have frequented for decades.

The only thing left of Bar K, once hotshots control the raging fire, is a pile of cinders and the wooden bear statue that once pointed to the parking lot. When Tarter reads of the news online, she sends an email to the SETI Institute staff with the subject line "the last milkshake."

The fires act as an uncomfortable reminder that our planet doesn't always act like it wants to be inhabited. It changes all the time, irreversibly turning landscapes we love—the ones where we meet our spouses and stake out our careers—into carbon.

❖

When Tarter took that first fortuitous trip to Hat Creek with Welch and Billingham, she had not yet finished her doctorate. She was still a graduate student, just starting to do SETI in her spare time. Her actual work—her doctoral dissertation—was meandering. She had studied everything from brown dwarfs to the esoteric Sunyaev-Zeldovich effect to instrumental analysis. Joe Silk, her advisor, just kept thinking of new problems for her to solve. Every time she nearly finished investigating one of his questions, he'd say, "What about this? Why don't you write something about this?"

It's a PhD student's worst nightmare—the kind where your teeth fall out, and you free-fall naked into the exam room of a class you didn't know you'd enrolled in. Unless a doctoral student's advisor deems their work "finished," they can't graduate. Tarter wanted to be done with school for all the normal reasons—autonomy, advancement, having been in school for 25 years of her 33-year life—but there was a more practical reason, too: she had already accepted a postdoctoral fellowship at NASA's Ames Research Center in Mountain View, California. But she had to finish her thesis before she could start work.

To finish, Tarter often came into the office after dark, a luggage carrier filled with computer punch cards trailing behind her. No one else kept vampire hours, so she was able to stuff the cards box by box into the computer without waiting in a line. Yes, computers used to have lines, because they cost more than houses and were nearly as large, so many people at the same institution had to share. Yet even today, there are still lines and applications for use time on super-computers like the one at Argonne National Laboratory in Illinois.

As her numbers crunched themselves, she sat in the empty Berkeley building, chastising herself for careless errors that negated some of the computer runs, requiring her to repunch the cards and then rerun them. But while some of this arrangement was for the benefit of her career, some, she had to be honest, was just about avoiding Bruce. Tarter "was getting more intoxicated about prospects of being a professional scientist and on [her] own."

"The wife thing was in my plans," she says, "but it didn't take star billing." She no longer felt that her and Bruce's bond was with each other, but primarily with their daughter, Shana, who held them together like the middle of a Venn diagram.

Bruce says their lives just became more and more separate, starting at that American Astronomical Society meeting in Puerto Rico, where she had had too much fun and cut herself on coral. Tarter had never had full adult freedom; she was a student when she had met and then married him. Bruce had experienced the world before their marriage; he had a real job. He acted distant; he was distant.

As Tarter grew up, Bruce says, they grew apart. "She had gone through undergrad early, almost without experiencing the things people normally do—parties, ups and downs, having a social world," he says. "She had never really done that."

Now, at Berkeley and in that wild "social revolution" Bruce referred to—which seems to have involved its fair share of swinging—Tarter was finally getting the chance. "We went to parties at Berkeley that I kept thinking would jeopardize my security clearance," says Bruce.

They lived on the same planet, and sometimes went the same places, but they inhabited different worlds. And the two began to separate those worlds even more. In the mornings, Tarter packed up her luggage cart, went home from the lab, and softly shook Shana awake. They ate breakfast together and then stood outside waiting for the school bus. Tarter zombie-walked back inside and went to sleep, the house hers alone, just like the computer. She woke up in time to meet the school bus upon its return. She felt as if no time had passed, like the two bus moments were stitched right next to each other, without a gaping loop of fabric between them, a wormhole through space-time.

After Shana settled in for the night and Bruce returned to watch over her, Tarter rolled the luggage cart back to Campbell Hall. She became a kind of wild animal that lived mostly to fulfill its basic needs, where "basic needs" were "write a 300-page scientific document and avoid my husband."

In the small moments she had free, she sorted through her shell collection—the one from her childhood visits to Manasota Key in Florida. The shells bore the marks of a different life from the one she had now. A life free and close to the ocean, full of big questions about the universe and her place in it and on this planet. She began building a table and placed each seashell carefully beneath a glass top. This table, she began to think, would look very nice in a place of her own.

❖

Despite marital tension, Tarter and Bruce hosted their annual gathering. The first Saturday of every May—even the Mays like this one, when their house felt more like a stage for performance than a home, to both of them—Tarter and Bruce held a Kentucky Derby party. It was part of Bruce's Southern-gentleman heritage.

As Tarter muddled the mint for juleps, she thought of Mika Salpeter. Her impeccable hosting. Her equal and companionable marriage with Ed. Tarter channeled Mika as she decorated the apartment with derby hats, rose wreaths, and horseshoes. She had gotten up at 5 A.M. to put the roast beef in the oven at the exact bacteria-killing temperature. By mealtime, the meat would equilibrate with the oven air and emerge perfectly rare. She cooked grits and wrapped 40 plastic glasses in aluminum foil, faux sterling cups.

As the guests arrived, Tarter smiled over their heads at Welch, put Joe Silk's perfectly rare roast beef in the broiler when he said it wasn't done enough for him, and mingled with all her guests. Welch smiled back.

❖

When December came around, Tarter and Bruce were still living in the same house. They sat in a room once a week with a marriage counselor. It felt clear, to Tarter, that this was the end. Tarter told Bruce she would not go to Christmas with his family but would stay and work. This abandonment was the first Shana knew of the impending divorce, her first experience of the split it would make in her family, and her first positioning as the only thing in the middle of the mother-father Venn diagram. She was not happy with that lot.

Not long after, Tarter began paging through the classifieds and walking past for-rent signs, imagining what life would be like with Shana in each apartment. She finally found a place on Virginia Street in Berkeley, a small two-bedroom flat that opened onto the backyard of the house.

"What I remember is that she just sort of delivered a fait accompli at some day, some time," says Bruce. "She said she had rented an apartment on Virginia Street."

And that was that. Their talk shifted to joint custody, separation. The terms felt like a foreign language, but one they were learning well, through immersion.

Days after their move, Shana set up an Army-surplus tent at Tarter's to match the playhouse that sat in the backyard of Bruce's house. Tarter rented an upright piano so Shana could continue her lessons. Tarter stacked unpainted particle board furniture, pre-Ikea, into desks and cubes and shelves—a whole identity to be arranged and rearranged against the walls. The glass-top shell table, finished, held a place of honor in the kitchen.

"I've always regarded it as an amiable divorce," says Bruce, "if unexpected from my coordinate system."

After all, he adds, she kept his name, and his subsequent ex-wives have not.

❖

After Tarter and Bruce split, Tarter began pursuing Welch. As Tarter speaks of this today, she pauses in the retelling and looks out at San Francisco Bay, where the fog is coming in and covering everything outside the house. Welch sits at the dining room table—an Italian contraption that expands such that no matter how many piles of paper sit on top, they can always undo a latch to make room for dinner. During one recent meal, the green beans came out overdone because Tarter and Welch had to finish a debate about single-dish radio telescopes. When dinners are over, they often dance to samba music in the living room of whatever house they are in, regardless of who else is present. They discuss the merits of the bossa nova artists that come up on Pandora and argue about whether they once saw this or that musician at this or that Berkeley club that closed in either the 80s or the 90s. They seem covalently bonded.

Welch, in the early days of their courtship, began to come over to Tarter's new room of her own in the evenings. One night, Tarter forgot to lock her bedroom door, and Shana walked into her mother's room to find only Welch, lounging on the floor mattress like a college student.

Tarter, hearing the shock, ran to the room.

"What's Jack doing here?" Shana asked.

But Jack was already *not* there, having launched himself out the window.

"Jack is going to be in our lives a lot more often now," Tarter explained to Shana, which was as uncomplicated as she could make it.

❖

Life was shaping up—an apartment, a daughter, Jack. Soon, maybe soon a career instead of school. But how soon? Silk, her doctoral advisor, continued to pull projects out of his sleeve like carnival scarves. One morning, despairing to think of living in this liminal state forever, she leaned against the wall of Campbell Hall and just cried. Ivan King, who had made the unfortunate comment about women on Tarter's first day of classes, saw her before she saw him.

"What's the problem?" he asked.

She wiped the tears away and stood up straight. "My time's almost up," she said, "and Joe keeps saying, 'Won't you write something else?'"

He wasn't requesting more and more maliciously—there were just so many questions to which he wanted answers, and here was this person whose job it could be to go investigate them! "I was one of his first grad students," says Tarter. "He had not yet internalized any obligation to guide the path of his students into their future careers. He never thought about the fact that I needed to move on."

King's expression loosened. He had had a friend during his own graduate school days who couldn't complete his thesis and move on to the real world, and felt like he would be trapped in limbo forever. Seeing no other way out, he killed himself. That's not an uncommon feeling, although it's a less common action. A 2006 study conducted by the University of California found that 60 percent of graduate students felt "overwhelmed, exhausted, hopeless, sad, or depressed nearly all the time." And 10 percent had thought about committing suicide in the past year. Tarter seemed so shattered that King feared she might be getting to that point.

"Come with me," he said. Holding her elbow, he marched them into Silk's office. He looked from Silk to Tarter and from Tarter to Silk. He pointed at the thin line on Tarter's paper where Silk's signature meant the difference, to Tarter, between this life and a different one.

"Sign," he commanded Silk.

He did.

❖

Now able to take a giant leap into the next phase of her life, Tarter and her friend Susan Lea, who also had an Ames postdoc, carpooled down I-880, across the 1.6-mile Dumbarton Bridge across San Francisco Bay, down the 101, and finally into Mountain View. They waved to the guard as they passed through NASA's security gates. They breezed into their offices and deconstructed the universe.

"We were hot shit," Tarter says. "We were handling full-time research careers, complicated family lives, and other avocations. The world should just come to expect this from women." Affirmative action, she thought back then, was totally unnecessary. Women were finally making it through the pipeline—in a funneled fashion, sure, but that would change. When the world saw how good they were, that world would say, "Oh, there's no problem! Women are fine. Bring 'em on."

But she had forgotten the political context of her own upstream swim. When she was in high school, she leaned against her locker listening to Sputnik *beep beep beep* Soviet dominance over the loudspeaker. The 1958 National Defense Education Act gave more than $1 billion to this teaching overhaul. And so Sputnik gave American scientists a chance they'd longed for: to rejigger curricula, shifting them toward basic research and adding innovations like hands-on labs (both the boon and bane of every high-schooler's existence, still today). The United States was so far behind that if catching up meant some scientists had to be women, so be it.

That attitude had withered by the time Tarter was a postdoc and the United States had climbed back toward the top. And while women of Tarter's day were perhaps more supported in their secondary-school years, the bolstering didn't yet apply (and still doesn't) to later-year complications. During Tarter's postdoc, the American Association of University Women was trying to remedy part of that problem. They invited her to come to Washington, DC, to attend a forum about a new policy. Congress was considering legislation to fund training for women who had left the scientific workforce to raise children but wanted to return. Tarter believed the cause was good, and she agreed to attend. But when a form arrived asking her who she would like to share her hotel room with, Tarter wrote back, "When somebody invites my husband to a meeting, they don't ask him to share a room."

When she arrived in DC, she walked into a room of 80 women. Eighty female scientists. Seventy-nine other people who were also hot shit. She had never been in a room with that many women, period, let alone that many with PhDs. Like a convention of people who'd spent time in solitary confinement, the women bonded over their shared isolation: living in a world that didn't quite know what to do with them—the geologists, she thought, had it the worst, as they were routinely denied fieldwork because the professors said their wives were jealous.

These women began investigating what they had in common besides male bosses with wives. They were all competitors, they found, all aggressive and gregarious (Jill's fifth-grade report cards, for instance, reported that "Jill talks too much," and then "Jill still talks too much." She had taken first place in county government day, but not because she was into politics—because she was into winning.) And while those inclined to exercise may have, had they lived in the 1990s, captained lacrosse teams, they were all former cheerleaders and drum majorettes, the only sports available.

But most interesting to Tarter was that many of these accomplished women had lost their fathers at early ages. Statistically, only 7 percent of kids have dads die before they're 20 years old. And although correlation doesn't equal causation, the women set out to find a cause. Why would parental death lead them to high-powered science careers? Because grief requires distraction? Because they learned to be independent? Because 50 percent fewer imposed societal expectations held them back? Because they were used to feeling detached and alone?

No, Tarter tells me. "It's carpe diem," she says. They knew if they wanted to do something, they should do it—right now. They wanted to do science. So they did it. Right then. Without caring about what other people thought. Tarter felt a kinship she'd never felt before.

She doesn't express regret often, taking a "mistakes are lessons from which you can learn" attitude. But she does seem chagrined by her attitude toward other women before this forum, an event that she says changed her completely.

"I got to realize how much I'd actually bought in to being one of the boys," she says. Particularly back then, to succeed and be taken seriously, women couldn't always be themselves, lest differences from male leadership encourage the male leadership to use phrases like *overly emotional* or *weak*.

Harp explains it this way: "She's developed this way of being, this way of dealing with people to sort of maximize her effectiveness. It works out better if she lets people think that she's tough. People pay more attention to her." The extent to which that is nurture as opposed to nature is impossible to know. But the tough shell is just that—a shell. After the first few interactions, or outside of business situations, it cracks. And she has learned to be nicer to others like her.

"Back then, I wasn't very kind to the few other women scientists that I'd met. Because I thought, 'Oh, we can all do it; nobody needs support,'" she explains. She may have even been tougher on those women, unconsciously. "It made me think about being kind and supportive to other women. And so if there has ever been an opportunity to hire a woman, I have."

Lindsay, Tarter's 2014 summer intern, is an example. She's more than 10 years older than most of the other students, who are barely allowed to drink the Coors Lights they buy at the store. Lindsay, on the other hand, went back to school for astronomy after working in television for years. She abandoned a career full of sitcom scripts and HD cameras in favor of a career full of . . . well . . . computer scripts and CCD cameras. It seems like a strange move: why school at all, why astronomy, and why this alien-based internship?

"I thought the only way to do the things I saw in movies was to work on the scripts," she says. "But then I realized all the things I like in movies are about science."

She could do science without the rat race of Los Angeles, the late payments and scrambling for gigs. She mentions *Back to the Future* and its "science-based" DeLorean. And then she smiles wistfully as she recalls the movie *Contact*. It inspired her.

"Dr. Tarter is basically the most famous female scientist," she says.

The other interns express similar sentiments. During their Hat Creek trip, they got into a debate, while eating sandwiches on top of a volcano, about the pronunciation of one particularly bright star: Vega (coincidentally, the star from which the extraterrestrial message comes in *Contact*). It's Vee-guh. No, it's Vay-guh. Back and forth. A few minutes into the argument, a quiet student on the outer edge of the circle spoke up.

"Dr. Tarter says 'Vay-guh,'" she offered.

The matter was settled.

❖

After their marital separation in 1975, Tarter and Bruce wanted to co-parent—with joint custody, as equals. But that was almost unheard of back then. So the couple remained married and shared custody for a year, while living apart, to show the confused judge empirical evidence that they could both be good parents, alone together.

They shuttled Shana back and forth between houses in Berkeley and Danville, sending her to a private school midway between. Shana had two beds, two sets of school uniforms, two toothbrushes, and after a while two dental retainers. Mom and Dad agreed: No good cop, bad cop.

Tarter took Tuesdays off to make a midweek weekend. They traveled to the Oakland and San Francisco zoos; the California Academy of Sciences; the Alexander Lindsay Junior Museum; the beaches at Point Reyes, Muir Beach, Alameda, and Point Richmond. Always, always ice cream cones.

When your life gets sliced into pieces, though, a scar stays even if you don't bleed out. The world wasn't the way Shana wanted it to be, and she was not in charge of it. But Shana healed herself well, in part, because of gymnastics. Here was an intact portion of the world that she could control, and the fact that the rest of her world was screwed up didn't matter.

"That was the place to be," Shana says of the gymnasium.

After Tarter and Bruce's empirical demonstration of co-parenting, the judge considered the evidence and concluded he should grant joint custody.

❖

Tarter's postdoc research at NASA's Ames Research Center, like her PhD work, dealt with brown dwarfs and what was then called the Space Infrared Telescope Facility, which became the Spitzer Space Telescope (it launched in 2003).

Still, Project Cyclops stared her down. She became obsessed with the idea of finding out whether humans are alone. She could not ignore the tingling it had left on the back of her neck—the feeling that she should turn around and look back. So when she heard that John Billingham—the NASA manager whom she and Bowyer had whisked to Hat Creek—was hosting a series of SETI workshops at NASA Ames, she attended the first one. The workshops would later come together into a NASA report called "The Search for Extraterrestrial Intelligence."

She walked up to Billingham and reintroduced herself. "I have more than forty hours a week," she told him. "I'd really like to get involved with SETI. What can I do to help?"

He nodded. As someone who first studied medicine, then designed spacesuits, and then swung to SETI, he understood the pull of the search. It was strong regardless of whether you had a full-time job, a new boyfriend, and a daughter. He invited her to come to his small Interstellar Communications Committee meetings, where the scientists asked questions like "Do we know what the sky looks like naturally, without alien broadcasts? How would we recognize an unnatural signal?" and "Do other stars have planets?" These were exactly the kind of questions whose uncertainties paradoxically set her mind at ease.

❖

"I was getting to know Jack and airplanes," Tarter says of the months after her marriage ended, when her SETI work was truly beginning. They began to see each other and hop in planes together for unofficial, non-alien purposes. They spent Sunday brunches at the house they soon began renting together. They flew to spontaneous picnic spots on the Baja coast, where they camped and watched the Milky Way. They found a fly-in-only resort in Oregon. They took afternoon airborne jaunts to Nut Tree, California's first major road stop, midway between San Francisco and Sacramento.

Soon, they took their first longer trip together—to the International Astronomical Union meeting in Grenoble, France. What happened there—a little wine, a lot of talking about thermonuclear reactions and spectral indices—isn't that memorable. But their return was. When Tarter turned in her travel expense forms to the National Research Council, she wrote, "The other half of the hotel room is covered by Jack Welch." That did not compute to the puritan council. But NASA higher-up and Tarter's friend David Black, who later had a brief stint as the CEO of the SETI Institute, pushed her paperwork through the system.

He told the council, "It's okay. They share a room every night."

❖

Tarter and Jack soon quit renting and purchased another house in the Berkeley Hills. Being there and looking across the water is like living in a lost jungle tribe and seeing a distant civilization you're not a part of—San Francisco—but then still being able to walk to the farmers market. So when they saw this house in the hills for sale, they made an offer without even going inside. That home, where they still live today, has cedar siding and sits on a one-and-a-half-lane road that cars have to bushwhack through. A second-story entrance is built into the hillside, with the deck 50 feet above the ground and redwoods threatening to block the view.

When you're standing outside the house at street level, it seems unassuming, with brownish paint and a wreath on the door. It looks like someday the trees might take it over. It's dark. But if you ring the bell, Welch will rise from the dining room table, wander over to the door, and open it. The light inside surprises your eyes. The whole far wall is windows, two stories high. Framed pictures cover the few walls that aren't windows—a Picasso sketch, a *People* magazine photo of a younger Jill and Jack, framed programs from Jack's daughter Jeanette's bass performances, a pixelated version of Hokusai's *The Great Wave*, titled *The Wave of the Future*. To sit in one of the ivory leather armchairs that faces the bay is to take a seat in the bridge of starship *Enterprise*.

In December 2014, Shana visited this house. She flew in from Lander, Wyoming, the tiny mountain town where the National Outdoor Leadership School is based and where she is the assistant director of the Wilderness Medicine Institute. Lander is the kind of place where everyone understands if you need to leave work at 1 P.M. for a 12-mile hike followed by some V5 bouldering to clear your head. She's visiting to help Tarter dump the furniture in her old bedroom. It's the same rearrangeable particleboard furniture that filled Tarter's first apartment on Virginia Street. After they finish at the dump, Shana goes to visit Bruce in Danville, where he still lives. *Plus ça change . . .*

❖

Around the same time Tarter, Welch, and Shana sewed their lives together, Tarter's scientific identity began to shift, too. She was moving from astronomy to becoming a SETI scientist. She, along with astronomer Jeff Cuzzi, flew to DC and met up with astronomer Tom Clark. The three then embarked on a long hairpin drive from DC to Green Bank, West Virginia. That scientific outpost, where Frank Drake performed the first SETI search, had updated its hardware in the decades since: Engineers had built a 300-foot telescope in 1962 (just after Drake's experiment). The dish tilted in just one direction, on an arc like the swinging ship ride at carnivals. From the edges of the dish, two arms reached toward the middle and touched at the top, where the receiver lived. That receiver turned the signals from radio waves into electricity, which carried coded within it information about the frequency of the radio waves from the cosmos and their strength. Cables then routed the signal down the arm to the control room, a squat brick building underneath the dish. The engineers designed the telescope's supporting structure and mesh surface to last only 10 years, just enough time to figure out what they should construct next.

The first night of their experiment, Tarter, Clark, and Cuzzi gathered in that control room. With its gray panels full of toggle switches, incandescent light-up buttons, and tuning knobs, the building resembled a cockpit. The team's flight plan was this: Figure out what narrow, station-like radio signals came naturally from the sky—from astronomical sources like supernovae and black holes. The scientists would have to be able to identify those and toss them aside during SETI searches, since they would be looking for similar signals that came instead from biological beings. How could they tell a molecular cloud from an intelligent civilization?

In the process, it was possible—always possible—that they would *find* that civilization, or even *those* civilizations. Searching for extraterrestrial civilizations is like waiting for the cable guy to come: you're always ready, on edge and dressed, even though he probably won't show within the anticipated window.

Clark knew how to use the 300-foot telescope from his days running a global observatory spread across the planet. A few years before, he had created a network of radio telescopes, linking them together to act like one telescope when he pointed them at the same object at the same time. It's kind of like how a bunch of fish team up into a school and swim together to look larger. For telescopes, the trick gives them better resolution: the telescope functions like one as wide as the biggest separation between any two antennas. So if you hook four 100-meter-wide antennas together, and the closest one is right next to you but the farthest one is 100 miles away, together they give the resolution of a telescope 100 miles wide.

In the pre-Internet days, linking the telescopes was not easy. It required recording all the data on magnetic tapes, with atomically precise time markers. Then, the tapes were synced with each other, like sound engineers line up actors' video mouths with their audio words. Tarter, Cuzzi, and Clark used these same magnetic tapes to perform this Green Bank SETI experiment. The exercise was aerobic.

Clark held up one of the rolls, 10 inches across and heavy as a barbell. He lifted it up onto the drive, fed its tape into the slot, and pressed go. The telescope couldn't just point at any object. It could move in a north-south line, up and down, pointing toward the horizon or up at the sky's apex. But it couldn't look east to west. It had to wait for objects to pass over as they rose and set. In their steady 12-hour journeys from one horizon to the other, they spent just three minutes strolling through in the telescope's view. During those 180 seconds, one tape took data. And the team needed to mount a second tape on a second machine to take data on what came next. Then, they had three minutes to go back to the first machine and set up another new tape. Ad infinitum (or so it seemed).

Tom Clark smiled. "Ping Pong," he called it.

The first tape of any set was "on source," meaning the telescope was pointed right at a specific object passing overhead. The second was "off source," meaning it was pointed at blank sky. If a signal appeared in both—if it showed up when the telescope couldn't actually see the star system in question—it wasn't "real." It didn't come from that star, but from Earth, a radar in Roanoke or a pickup's spark plugs. This is actually the hardest part of SETI: distinguishing ourselves from the other.

The scientists prepared to look at their list of candidate stars, bouncing back and forth between the tape reels. Tarter knew it was wise, even then, to keep her hope at some baseline level—too much crest led to too much fall. And the potential for being fooled, being wrong. These were the careful early days of a new relationship, when you don't want to betray your fervor.

So, stoically, they looked at her long list of sources. On. Off. On. Off. Tape. Tape. Tape. Tape. And then, coming from one of the star systems, a beacon appeared. It showed up in the on source.

Disappeared in the off. There in the on the next day. Gone in the off. Just what they were looking for.

By the third check-up, they were all a little wide-eyed. They watched it appear in the on-source observation. Yes. Good.

But when they tried to look at the off tape, they found it was corrupted. No results. They would have to spend a restless, no-REM-sleep night waiting for the source to rise again above the horizon.

Tarter allowed herself to wonder if the signal was "real."

She called up Billingham.

"What do we do if this turns out to be right?" she asked.

There was no protocol for who to tell what when and how, he said. She was on her own.

❖

On the fourth day, they pointed the telescope at the star. The signal showed up. Then, the sky moved and the telescope pointed away from the star, to the off-source position. And the signal came back—just as bright as ever. For a microsecond, Tarter felt the jolt of the discovery, almost as if it were pulsing through her. But then her brain intervened: *Remember, a real, right signal wouldn't show up in the off-source tape.* The telescope wasn't looking at the star system, so if the signal still remained, it had to be coming from somewhere else. It was likely earthly interference—not evidence of alien technology, but evidence of *human* technology. The scientists had found us—intelligent, technological, communicative hominids.

But what was this technology? And how had it mimicked a space source for so long?

It all comes back to the way the world turns. The physical day is actually 23 hours and 56 minutes long (the missing four minutes are what we make up for with leap years). For a star to travel from one spot in the sky all the way around Earth (from our perspective) and appear again in that same spot doesn't actually take a full 24 hours: It just takes 23 hours and 56 minutes.

Someone, Clark said, was turning on a piece of electronics at the same Earth time—around 8 A.M.—every day. The first three days, that didn't match up with the time the off-source location was overhead. But each day, the star's overhead appearance slid closer to 8 A.M. by four minutes. And on the fourth day, both times—the time they looked at the star and the time they looked off—fell after but near 8 A.M.

Eight in the morning, Clark realized, is when shifts change. Telescope operators get in their trucks and drive down the long NRAO road toward the lone highway, to the diner or to say goodbye to their kids before school. And when they reach the gate, they turn on their CB radios. *Ping.* An intelligent signal. On day 4, the CB was still within range when the off-source tape was recorded.

Despite the disappointment, it was on this trip that Tarter decided 32 was the perfect age—because it was her age—to become a radio astronomer.

"A very strange sort of radio astronomer," she clarifies.

A SETI radio astronomer.

❖

Twelve years later, in 1988, the 300-foot telescope the team used for this experiment was 16 years older than it was ever supposed to be. The National Radio Astronomy Observatory had built it as an interim solution, intending to tear it down and make something better within a decade. But money is always tight in the sciences, and when the 300-foot continued to work well, they just kept it around. Until they had no options.

On November 15, 1988, telescope operator Greg Monk helmed the controls when metallic sounds rang around him. The building shook, like an earthquake that came from above. He ran outside to see what was happening.

All around the control building, pieces of mesh and scaffolding made a pile of debris like ruins some future society might find. The

300-foot telescope had collapsed. Monk walked back inside to find that a support beam had pierced the toilet.

The next day, tabloid papers claimed extraterrestrial laser blasters had destroyed the telescope: ZAPPED! BY HOSTILE SPACE ALIENS, one headline claimed. If only! That may be the second most likely cause, but the first is metal fatigue. The telescope was just old and tired.

In Green Bank's modern-day gift shop, you can buy a tiny piece of the 300-foot's mesh surface, hot-glued to a wooden plaque with a picture of the telescope in its fresh-faced youth. The plaque lists birth date and death date, as if the photo belongs on a table at a funeral.

❖

Back in 1980, after her Green Bank trip, funerals weren't on Tarter's mind. But marriage was. She and Welch had both been granted sabbaticals, which they planned to spend in Europe. Passports and paychecks would play out more easily if Jill and Jack were Jill and Jack Tarter-Welch.

Shana didn't understand why her mother and Welch needed to get married. "Things are just fine the way they are," she said. It was just a piece of paper, and Welch lived with them already.

But besides the logistics, Tarter told her daughter, "It's a good idea. I like the idea."

Shana thought about that for a while, considering whether she also liked the idea, and whether she liked the idea of her mother acting on this idea that she liked.

"Just don't change your last name," Shana said. Her class roster ran long with hyphenated names that took the whole afternoon to pronounce. "I don't want to be one of those."

Tarter agreed. Besides, in journals, women who changed their names became marked with a scarlet-letter superscript, accompanied by a "formerly known as" footnote. "It felt like 'aka a criminal,'" Tarter says.

Tarter and Welch married on July 4, 1980—a day representing togethered independence, and a date neither of them would accidentally forget in the years to come. Welch flew them to Gualala, north of San Francisco Bay, for their honeymoon.

That year, she sewed Welch his first paisley shirt. An authentically Instagram-hued photo now shows the shirt hanging from that year's Christmas tree.

"Life," she says, looking at the picture in 2015. "Life."

She closes the photo album.

CHAPTER 6

THE POLITICS OF SCIENCE AND NEW PROJECTS

O n her first trip abroad, Jill Tarter, soon to be 37, stood next to the baggage carousel in the Charles de Gaulle Airport in Paris, waiting for her duffel bag. It was as heavy and over-stuffed as a couch. Next to her, Welch waited with a slim suitcase containing a single pair of pants, no shoes, and a few shirts. She admired him for knowing how to pack a suitcase worthy of the international terminal. It was worldly, she thought. The sense of sophistication was something she admired in Welch and had also found attractive in Bruce. It was something she wanted for herself.

It's hard to imagine, now, that Tarter was once provincial, maybe even a little rough. She attends the World Economic Forum; she has season tickets to the San Francisco Symphony and is genuinely familiar with the works of contemporary architects.

This striving for significance added to her excitement about a meeting that was set to occur at UC Berkeley in 1980. Eminent scientists like Carl Sagan and Philip Handler met at the university's Space Sciences Laboratory to discuss an idea that sounds almost as science fiction as aliens: the catastrophic events that cause mass extinctions, like the ones that slayed pseudosuchia (creepy pre-crocodiles) and dinosaurs, which occur with almost clockwork-like regularity.

Everyone began to seat themselves for the briefing, while those in charge debated whether recordings should be allowed. Like so many game-altering ideas in science, this concept of repeated occurrences of extinction-level grand events throughout the history of life on Earth could be a crazy fantasy—lifted out of dubious and scattered statistics—or it could change our conception of life on Earth. As officials readied the equipment, Billingham led Tarter toward a smiling man in a turtleneck.

"This is Carl Sagan," Billingham told Tarter, gesturing toward the floppy-haired man that everyone in America already knew for his role in the television show *Cosmos*. Sagan reached out his hand.

"And this," Billingham continued, "is Jill Tarter."

Before they could talk more, the meeting got started. Scientists Walter Alvarez and his father, Luis Alvarez, walked up to a podium and presented their half of the big announcement. Walter had been on a geological expedition in the canyons of Gubbio, Italy, looking at limestone. In it, he found a thin layer in between the regular rock strata, like the middle part of an Oreo cookie. Looking at the different layers of rock on our planet is like looking back in time, just as looking farther out in space is. The farther down you go into a canyon hike, for instance, the older the rock next to you is. Each layer was, at one point, on Earth's surface. So when a strange set of chemicals appears, you know something strange was happening on Earth's surface at whatever time that rock formed. And this layer of rock happened to be from the time when dinosaurs went extinct.

Walter brought a sample of the mineral sandwich to Luis, who had access to a mass spectrometer. A mass spectrometer is like a fingerprint reader for elements and molecules, revealing the chemical identity of a sample. When the results came back on the rock layer, they revealed a lot of an element called iridium—a shiny metal that's the second-densest in the universe, and one that doesn't appear often on Earth.

What could have left so much iridium? the Alvarezes asked each other.

The only thing that made sense, they eventually concluded, was a space invader—like an asteroid, a type of object known to be rich in iridium. If an asteroid had crash-landed, it would have brought its own chemistry and then splattered that chemistry into the sediment. When Walter and Louis figured out how long ago this collision would have happened, they came up with the answer to every kindergartner's favorite question: what killed the dinosaurs?

An asteroid, they concluded. This same asteroid that left this iridium.

Then, two other scientists—David Raup and Jack Sepkoski—took the stage. Billingham had funded their research into the extinction of sea animals, and they had made a morbid timeline of the mass deaths in Earth's oceanic past. Every 26 million years or so, as if on schedule, swarms of species disappeared.

The four scientists then put their work together into something more significant than the sum of the parts: extinction-level space rocks might hit Earth, like trains arriving at a station, every 26 million years.

The first time you hear it, it makes your spine feel cold. It makes the universe sound malicious, intentional—like something is *in charge*. But the scientists set out to find an astrophysical cause. The meeting's attendees threw ideas out fast and furious, demolishing most of them in the next sentence. Maybe an unknown planet (Planet X) or a nearby star (the Death Star) passed close to the outer solar system with some orbital regularity. Its gravity could perturb the rocks in the asteroid belt, flinging them to Earth. Or maybe as the solar system moves in its orbit around the galaxy's center, it moves in to denser pockets of space at regular intervals, and that extra mass does the same rock flinging.

Regardless, the report that came out of the event concluded that "the work of the Alvarez group has emphasized something that should have been recognized earlier: The Earth is not alone!" That's something Tarter had thought about in terms of planets like Earth and intelligent beings like humans, but she had not quite considered how much extraterrestrial factors affect the life that already exists here. Those interwoven threads became part of her worldview: we are *part* of the universe, not beings simply peering out from a bubble and into *a* universe.

After the meeting was over, Tarter walked with Sagan back to the parking lot.

"What would be the level of light," he wondered out loud, almost as if to himself, "if one were able to stroll on the surface of Venus, rather than in the Berkeley Hills?"

Tarter struggled to come up with a plausible answer, but she needn't have bothered. Sagan responded to his own question: about as bright as an overcast day on Earth.

Sagan was always "on," Tarter thought, always performing for a bigger audience than the one present. And she was right: When episode 4 of Sagan's *Cosmos* TV series later aired in October 1980, a showing that Tarter and Welch watched with their blended family, Sagan asked the TV viewers the exact same question.

And he never forgot Tarter. Although he did not himself participate in much SETI, he would become a public advocate for and educator about what he considered a vital human question, as well as an admirer of the person—Tarter—who did much of the day-to-day work and big-picture strategizing about how humans could investigate that question, rather than just talk about it.

❖

Soon after the meeting, Tarter and Welch took their sabbatical year in Europe. In Germany, they lived in an apartment underneath the bowl of the Effelsberg Telescope, yet another radio antenna. Shana, as well as Jeanette and Leslie, Welch's daughters from his previous marriage, hated it. The only good thing about the city of Bonn, they thought, was that the McDonald's sold beer.

Better for everyone was Paris, where they lived at no. 1 Place Paul-Painlevé, at the junction of Rue Saint-Jacques and Boulevard Saint-Germain, across from the front door of the Musée de Cluny. The flat had 15-foot-high ceilings and parquet flooring with ruts from hundreds of years of familial footsteps. Each morning, Tarter and Welch bounced down the stairs, past the outdoor pipes that brought water up because the place had been built before plumbing, and out onto the Rues. They jogged through le Jardin du Luxembourg. At night, they went to Alliance Française for language lessons.

But money from the university didn't always appear when it was supposed to. There were no ATMs, and no automatically calculated exchange rates. So when their wire transfers from Berkeley arrived in the Old World, the money went into a French bank account, where it waited for months to be joined by the francs supplied by their French hosts. Near the end of their sojourn they were scarily cash poor. Every day, Welch went to their bank to inquire about the payment.

"Has it arrived?" he asked the teller.

"*Pas de tout*," the teller responded, every day.

But at the last minute, the teller responded, *"Oui!"*

Tarter sewed the boatload of paper money to the inside of Welch's jacket for their upcoming trip home, home ec talents meeting smuggler skills. Except for sleeping, Welch didn't take that jacket off until they were back in the States.

They once again loaded up their Saab, baggage in the trunk and on the roof, to ride the hovercraft that would take them across the Channel. Shana lifted her suitcase to her mother, who placed it next to the others and tied them all to the top. Shana's luggage was filled with new European clothes. She was much more sophisticated,

137

having spent this year acquiring fine taste in fashion, than her mother had been at her age, despite the fact that Shana now claims her mother is the fashionable one and the good-clothing gene is recessive.

But when they arrived at the port, Shana's shopping suitcase was gone. They all imagined some field in the quaint-cottage countryside, now covered in silky blouses.

"Everybody back in the car," Tarter said.

Everyone drove back to Paris; everyone scoured the shoulder the whole way. But they found no trace of the clothes. When they finally got to London, Tarter liberated some of the cash from Welch's jacket and sent Shana on a shopping spree.

Tarter calls this mishap their "second demerit as parents."

The first was years earlier, when a 10-year-old Shana interrupted the dinner conversation, which as usual revolved around astronomical mysteries and scientific politics. Out of the blue sky, Shana declared, "I want to be a shopkeeper."

"How did you decide that?" Tarter asked. "And what do you want to sell?"

"I don't know what I will sell," Shana said, "but I want to be a shopkeeper because you close at five and leave all your work behind."

The oblique, yet pointed, rebuke clearly still stung. Still, Tarter's colleagues, and her daughter, too, agree that she did put priority on her family, perhaps at the expense of the sleep she doesn't get. "Although all we saw was the working side of Jill, she has this softer side that isn't always visible to us," says Harp. "You could tell by the way she talked about her daughter and granddaughter that she really maintained a life outside of all the work that she did."

❖

Tarter and Welch didn't yet know it, but they were going to return to trouble when they got back to the States. A few years earlier, in 1978, a senator from Wisconsin—William Proxmire—had discovered

NASA's SETI program, a fledgling thing that the agency had just begun funding in 1975. Called the Microwave Observing Project (MOP), it was under the leadership of Billingham. MOP, at that point, was just design studies focused on the nuts and bolts of engineering the technology but not yet actually building that technology. A year later, NASA formed an official SETI department at Ames with a complement at the Jet Propulsion Laboratory in Pasadena. Proxmire didn't like it.

Once a month, Senator Proxmire skewered one publicly financed project that he believed was a waste of taxpayers' hard-earned coin. Upon these unlucky projects, he bestowed his Golden Fleece Award. The name alludes to both the verb *to fleece*, meaning "to con or overcharge," and the chivalrous 30-knight medieval group called the Order of the Golden Fleece. Every month from 1975 to 1988, he sent a press release giving one group this dubious honor.

He bestowed his first prize on the National Science Foundation, which had shelled out $84,000 to study why people fall in love. "I object to this not only because no one—not even the National Science Foundation—can argue that falling in love is a science," he said, "not only because I'm sure that even if they spend $84 million or $84 billion they wouldn't get an answer that anyone would believe. I'm also against it because I don't want the answer. I believe that 200 million other Americans want to leave some things in life a mystery, and right on top of the things we don't want to know is why a man falls in love with a woman and vice versa."

Proxmire, who looked like a bald hawk, was fastidious and frugal in all aspects of his life. He never missed a roll-call vote. He gave 3,000 separate speeches supporting an anti-genocide treaty. He jogged 10 miles every day and even wrote an evangelistic book called *You Can Do It: Senator Proxmire's Exercise, Diet, and Relaxation Plan*. And while he did get hair transplants and a facelift, he refused campaign donations, paid his own travel costs, and spent less than $200 on self-promotion each election cycle. No one was ever going to give *him* a Golden Fleece Award.

In 1978, he set his budget-slashing eye on NASA's fledgling SETI program.

> *I am giving my Golden Fleece of the Month award for February to the National Aeronautics and Space Administration, which, riding the wave of popular enthusiasm for* Star Wars *and* Close Encounters of the Third Kind, *is proposing to spend $14 to $15 million over the next seven years to try to find intelligent life in outer space. In my view, this project should be postponed for a few million light years . . . While theoretically possible, there is now not a scintilla of evidence that life beyond our own solar system exists. Yet NASA officials indicate that the study is predicated on the assumption that intelligent extra-terrestrial beings are out there trying to communicate with scientists here on Earth. If NASA has its way, this spending will go forward at a time when people here on Earth—Arabs and Israelis, Greeks and Turks, the United States and the Soviet Union, to name a few—are having a great difficulty in communicating with each other . . . At a time when the country is faced with a $61 billion budget deficit, the attempt to detect radio waves from solar systems should be postponed until right after the federal budget is balanced and income and social security taxes are reduced to zero.*

First, it bears pointing out that a light-year is a measurement of distance, not time. Second, these are not unique complaints. Today—and in reference to basic research on everything from supernovae to shrimp—practical people say their tax dollars would better be spent directly improving human lives. There's some merit to that, on its face. But Ed Catmull, in his book *Creativity, Inc.*, wrote a smart rebuttal to that sentiment and to the Golden Fleece Awards generally:

> *The truth is, if you fund thousands of research projects every year, some will have obvious, measurable, positive impacts,*

*and others will go nowhere. We aren't very good at pre-
dicting the future—that's a given—and yet the Golden
Fleece Awards tacitly implied that researchers should know
before they do their research whether or not the results of that
research would have value.*

But the SETI scientists, in their disgrace, were in good company. A great deal of legitimate, respectable scientific research—
Department of Justice investigations into why prisoners desire escape
and studies of the sex life of screw worms, which ultimately led to the
cattle parasite's eradication—had been similarly roasted. It wasn't a
career or project ender for anyone. While scientists sometimes sued
Proxmire for libel or felt head-hanging angst, the SETI team was
more used to public ridicule than most scientists, then and now.
After all, they *were* searching for smart aliens. Tarter thought the
stigma would disappear the next month when Proxmire moved on
to someone else's science.

Frank Drake, SETI's founding father, was a bit angrier: he
nominated Proxmire for membership in the Flat Earth Society, in
an article published in *The Scientist*. "When Christopher Columbus
left Spain, there was no evidence the New World existed," Drake
told the reporter, "let alone Wisconsin."

The award didn't ruin them. But a few years later, in 1981,
Proxmire decided to wage full-on war against SETI. He read an
article in the magazine *Reason* called "NASA Flimflams Congress."
According to this piece, "NASA officials publicly maintain that no
SETI activities continue at all today, except perhaps for a very few
low-level investigations that are not part of any organized plan. In
my first inquiries to NASA I was told outright that no SETI program exists, and *Reason* editors checking up on my story were told
the same things."

The author, Robert Sheaffer, then offered (true) evidence to the
contrary. His editor, Marty Zupan, wrote in a sidebar that Donald
DeVincenzi, head of Ames's Exobiology division said that "Ames and

JPL were doing a little work on how it would be done if it were to be done," but in the parlance of the agency, there was "no SETI project."

NASA *did* have a SETI program—the same MOP design study program they had had for years—but they hadn't been trying to hide it, insists Tarter, just as DeVincenzi implied.

Like most stories, this truth seems to be midway between these two. NASA probably didn't want to broadcast its alien activities, and so they may have been semantically slippery about their word choices. "No SETI project" may have been an evasive way to say "no *active* search is going on" while alluding to the R&D activities, which *were* for a future SETI search.

When Congress met to discuss the 1982 budget, Proxmire stormed in, wielding this article. NASA obviously had not taken SETI's Golden Fleece Award seriously. He and his fellow congresspeople sliced SETI entirely from that year's projects.

❖

Crushed, Tarter and Billingham sat down together to write an official termination plan, detailing how to dismantle their dreams piece by piece. It was like being shot in the foot and then being told to amputate it yourself.

But Billingham carefully, cleverly found unspent money from the previous fiscal year that he could distribute throughout the next 12 months—so this would be merely a lean year for the team, not a famine. He kept them and their brand-new prototype equipment alive. And in the meantime, SETI's biggest celebrity—Carl Sagan, who thought and spoke a lot about SETI—went to Proxmire's office to chat.

Sagan believed that like most other politicians in the 1980s, Proxmire would respond to Cold War rhetoric. If we found an alien signal, Sagan told him, that would be an indication that technological civilizations like ours can survive without nuking each other out of existence. In a world where adults built air raid shelters and kids hid under desks for bomb drills, the logic resonated. It's something SETI scientists still say, although the specific threats to humanity's survival they might reference have changed. A long-lived alien civilization means there's hope that we can turn around or adapt to climate change and not exhaust all our resources with our huge population—that's still a hard hitter today.

Perhaps it was Sagan's charisma, or perhaps Proxmire really did care about cosmic longevity. Regardless, Proxmire relented and agreed to restore SETI's budget in 1983. Sixty-nine prominent scientists from 12 countries also signed a letter, published in *Science*, saying the world should begin "organization of a coordinated, worldwide and systematic search for extraterrestrial intelligence."

And so in 1983 NASA reestablished its SETI program, with Billingham as team leader. "But NASA didn't just try to slide it in again," Tarter says. "They went back in and said, 'Hello, we're not cutting this. This is something we really want to do, and here are all our justifications.'" This is the biggest, most human question humans have ever investigated; if we don't seek we'll never find; we finally have the capability. It was the *Cyclops Report* all over again.

NASA requested $1.5 million each year for five years to continue work on a multichannel spectrum analyzer (MCSA), the prototype SETI instrument Tarter and the other NASA types like John Reykjalin had been working on for MOP. It would detect radio waves and then split them into tiny channels, or stations, as small as 1 hertz each. Automated software would then search each station for anything more than static. They planned to connect the MCSA to an existing set of radio telescopes in remote regions of Puerto Rico, Australia, France, and West Virginia, as well as a set of smaller NASA-owned telescopes called the Deep Space Network. These radio telescopes lived in near Madrid, Spain; Tidbinbilla, Australia; and Goldstone, California. Normally, NASA used these antennas to send commands to and receive data from spacecraft, like *Voyager*, launched in 1977, then somewhere between Jupiter and Saturn. Now, however, the network would receive data from—well, who knows what, or who.

❖

Before the SETI scientists could find smart-looking signals from space, though, they needed to understand the strange signals the universe produces naturally, and the signals humans put out with their satellites, sitcoms, and station wagons. SETI needs to be able to distinguish these natural and human-made radio waves from the ones that might come from extraterrestrial beings. As the SETI team began to plan an actual search for extraterrestrial intelligence, Tarter focused on finding the narrowest-band radio waves that came from astronomical objects, not biological beings. Most very narrow signals that we know of are synthetic—made by us. So SETI scientists thought extraterrestrials might compress their signals, too. But to figure out how concentrated a signal had to be to look "smart," Tarter needed to know how concentrated nature made its own signals.

Every color we see is actually just a particular wavelength of light. Red is 650 nanometers; yellow is 580 nanometers. Our sun looks yellow, but it doesn't just emit yellow light. It shines across a wide spectrum of colors, with yellowish being the brightest. Lasers, on the other hand, send out "coherent" light, much more focused around just one wavelength. Space has its own lasers (although space chose to call them masers, which stands for microwave amplification by stimulated emission of radiation) made of radio waves, which are the most coherent radio sources in the universe. You can't see them with your eyes, because your eyes can't detect radio waves. But if your eyes were radio telescopes, the masers would look just like lasers. For one figure who has been important to SETI and is famously depicted in fictional form in the movie *Contact*, radio waves *are* like visible light waves. Kent Cullers, the blind astronomer who worked on SETI beginning in 1985 and contributed key signal detection hardware and software, truly gets radio waves. "My sensory connection to the wider universe is not vision but radio waves," Cullers said in a SETI Institute interview. ". . . Because Braille can now represent mathematics and diagrams, not only the world but also the universe is open to blind people."

SETI's spectrometer, with which Cullers worked, could split an incoming radio signal into tinier channels than any that came before. The scientists could tell, for the first time, exactly *how* compressed masers' radio waves are. Then, in the future, if a SETI instrument picked up anything more compressed than that, the scientists would know the signal was either from technologically competent extraterrestrials who squish their signals like we do, from some undiscovered kind of celestial object, or from humans. They called this "defining the SETI sandbox." And, like curious kids, they wanted to play in it. But getting time on big telescopes was hard when you just wanted to test something and not do a more meaningful experiment. Usually, for national telescopes, a jury of scientist peers ranks written-up proposals and allocates the time to the worthiest projects. SETI didn't often make the cut.

NASA, wanting to help the project they had funded succeed, suggested Tarter use the JPL's radio frequency interference–detecting van for her project (yes, a van). It didn't have vinyl seats and Indian-print tapestries like other 1980s vans. Instead, engineers had stuffed it with scientific instruments that could create 65,536 radio channels as narrow as 300 hertz each. For comparison, an FM radio station spans 50,000 hertz.

One day, Tarter received a call from JPL astronomer Sam Gulkis: Jodrell Bank Observatory in England wanted to borrow the van. And if Tarter and her team would agree to shepherd it across the pond and show the British scientists how to use it, they could do SETI on the observatory's 250-foot Lovell Telescope. The offer seemed too good to be true. And as with $10 Coach handbags, the offer was, indeed, too good to be true, something Tarter wouldn't discover until she was already ensconced at the observatory.

❖

In England, Tarter found that while she loved tea, she hated tea time. During each ceremonious drinking session, she bounced her leg like she was about to lift off, eager to get back to work. She longed for the rhythmic hum of cryogenic dewars and cooling fans rather than the hum of conversation. The observatory's director didn't work much at all in the afternoons. Lovell gave public tours of his landscaping, the gardens on the telescope property, where he lived. Look what man hath wrought! The universe may be beyond mastery, filled with black holes and supergiant stars he could not touch, but he could master the terrestrial domain.

She shared the spectrometer with Gulkis and other scientists and engineers she hadn't met, and she also teamed up with Jim Cohen, a radio astronomer from Manchester, who was an expert on masers. Something seemed off, though. The engineers were always tinkering and fixing, although it seemed like nothing was wrong. Still, her and Cohen's search was going well, as they systematically narrowed the channel widths and found no radio signals as squished as 300 hertz. That meant SETI's future 1-hertz instruments would reliably turn up only synthetic signals, whether from extraterrestrials or Earthlings.

"This is great!" Tarter said to her JPL colleagues. "This is the first time we're doing SETI observations outside the country."

They nodded, nervously laughing and coughing into their hands. "Yep," they said. "It sure is a big moment for SETI."

She shrugged their strangeness away, chalking it up to the everyday awkwardness of engineers.

"And then I found out," she says, "it was all a sham."

For 21 years, Lovell had been obsessed with detecting radio communications from Soviet spacecraft (also known as spying). Although he had a whole observatory at his disposal, he had never been able to intercept the signal. Neither had the US government—which was also interested. When Lovell heard about SETI's spectrometer, he realized it was the perfect instrument to pick out the weak, narrow signal from a Soviet Venus probe called *Venera*. He could hijack its high-resolution images of the planet's surface. So together with scientists at JPL, he had dreamed up a scheme: they would tell the SETI scientists they could do their preliminary work at Jodrell Bank. They would let the SETI scientists hook up their instrument, they would tell them they were doing SETI, and then they would, behind the scenes, spend most of the time trying to find *Venera*. The whole team knew about the scheme—except for Tarter.

When she found out from Gulkis, she stormed from their meeting room. And then she stormed right back in. Her colleagues were surprised: they had assumed she was in on the secret, too.

"I thought we were doing SETI here, guys!" she said.

They went back to shrugging and looking askance.

Lovell, it turns out, did find what he was looking for. In 2011, the National Security Agency declassified a document titled "The Longest Search: The Story of the Twenty-One-Year Pursuit of the Soviet Deep Space Data Link, and How It Was Helped by the Search for Extraterrestrial Intelligence."

The report details the history of the hunt, including the SETI involvement. "The SETI specialists were given sanitized search parameters and limited feedback on results," the report says. And then the kill: "At 0635Z a teletypewriter at DEFSMAC clattered briefly with a crisp message: 'We have it.' The twenty-one-year search was over," it proclaims.

Ruse aside, Tarter still got her data: The most squished natural signals were indeed 300 hertz wide. These came from saturated hydroxyl (OH) masers, which shine from the vicinities of stars and from faraway galaxies that emit a lot of infrared light. But her hoorays were muted.

Despite Tarter's annoyance, she kept an end-of-stay appointment with Lovell. When she entered his office, she found him standing behind his desk staring at a coffee-table book. "He's looking at these pictures, and he's crying," she says. "Think of an elderly British lord of the realm, tears just streaming down his face."

The pictures were of Dresden, after World War II bombs turned it into a flat, fiery place.

"My sister tells me I should be ashamed," he told her. He had helped develop the radar technology that allowed the bombers to drop their payloads so accurately, killing 135,000 and decimating a once great city.

Tarter looked down at the pictures, at the horrible things humans can do to each other.

"I didn't experience that," she said, "but I think that's the way London would have looked if not for you."

He nodded and closed the hardcover.

"I never did make it to his garden," Tarter says now.

❖

When Tarter returned, she and two hundred other astronomers descended on Cornell University to celebrate Ed Salpeter's 60th birthday. Salpeter was a beloved figure in astronomy—a humble generalist and great skier with a good sense of humor, and the man who'd taught Tarter's first astronomy course. "I don't ever remember Ed working on a trivial problem," colleague Yervant Terzian said in Salpeter's Cornell University obituary. He was "always ready to look at new problems in new fields, and a young colleague quoted him as saying there were problems to be solved on backs of envelopes of various sizes," wrote astronomer Virginia Trimble for the American Astronomical Society. Many attendees said, at the celebration, that Salpeter taught them that scientific research should be a joyous experience (a lesson some scientists never learn). And indeed it was Salpeter's star formation course— full of the most interesting questions, Tarter had thought—that had spurred her on to her study of astronomy in the first place. Perhaps it was his zest, so contrasted with the joyless engineers' attitudes, as much as his source material that compelled Tarter and impelled her toward her current career.

At the party, thirteen scientists, including two Nobel Prize winners, gave the gathered partiers an update on the state of the universe, from black holes to dark matter to the scientific publication process (do astronomers know how to have a blast, or what?).

On the same trip, Sagan, whom Tarter now had seen at several scientific conferences and workshops on SETI, and his wife, Ann Druyan, invited Tarter and Welch to a cocktail party at their cliff-hanging home above the Ithaca gorges. The two had been married for just three years. Their love was so electrifying that a brain scan of Druyan contemplating Sagan is etched into a golden record placed aboard the *Voyager* spacecraft—a record meant for extraterrestrials to find, or not, and learn something about humanity through our images, astronomical plots, music, and neural patterns. In a picture from that same year, Sagan and Druyan's eyes both bore holes through the film—Sagan's with ingratiating smile lines, Druyan's with a 10,000-yard intensity.

"Carl's writing a science fiction book," Druyan whispered to Tarter.

Tarter rolled her eyes. Everybody knew (Sagan and the *New York Times* had made sure they knew) about his record-breaking publishing advance—$2 million.

"We think you'll recognize someone in it," Druyan continued.

"But I think you'll like her," added Sagan.

Sagan, she discovered, had modeled the book's main character on her.

"As long as she doesn't eat ice cream cones for lunch," Tarter told him, referencing a well-known guilty pleasure, "no one will think it's me."

Besides, she thought, how much "her" could it be? Sagan was just a famous guy she ran into a few times a year in professional settings. But then he sent her a pre-publication copy.

"I remember reading it, and it just felt so *familiar*, down to the death of my dad, an uneasy relationship with my mom, and the 1957 T-Bird, my fantasy car," Tarter says. "How did he get inside my head? How does he know?"

To be such a charismatic figure, Sagan likely had to understand the drives and motivations of those he interacted with. When Sagan looked at Tarter, it seemed he looked through her, X-raying to find out what her gears and switches looked like, what made her tick.

Later, moviemakers used Sagan's fictional alien-hunting character—whose name was Ellie Arroway—to imagine SETI's financial situation correctly. In *Contact*, the government cuts funding to Arroway's SETI project, forcing her to shrink her dreams. Then, they terminate the project, and Arroway has to travel the world in search of private donors to fund the project. In one climactic scene, she pitches the importance of SETI to a world-dominating tech company called Hadden Industries. The sharp-suited gatekeepers, CEO S. R. Hadden's employees, shake their heads. Her proposal sounds like "less like science and more like science fiction." Arroway explodes all over their sterile corporate boardroom.

"Science fiction," she repeats. "You're right, it's crazy."

She slams down her poster. "In fact, it's even worse than that," she continues. "It's nuts. You wanna hear something really nutty? I heard of a couple guys who wanna build something called an *air*plane. You know, you get people to go in and fly around like birds. It's ridiculous, right? And what about breaking the sound barrier, or rockets to the moon? Atomic energy, or a mission to Mars? Science fiction, right?

"Look, all I'm asking is for you to just have the tiniest bit of vision. You know, to just sit back for one minute and look at the big picture. To take a chance on something that just might end up being the most profoundly impactful moment for humanity, for the history . . . of history."

It's a speech not unlike those Tarter would later have to give to the Silicon Valley tycoons, when the government funding agencies terminated her own project. She, like Arroway, would have to travel—as she still does today—fundraising for her nutty-sounding scientific idea. But when the book *Contact* came out, SETI was still looking up. They'd just gotten their project reinstated by the government, after all.

"C'mon, Carl," Tarter once said to Sagan, who'd helped with the movie script. "You're going on about this, but we're fat, dumb, and happy. We're back within the fold at NASA. This time there's too much momentum; this time it's going to happen. I guess it's just more dramatic this way, huh?"

But, as they soon found out, Sagan was "maddeningly good" at looking inside things—not just people but also organizations and governments—seeing how they worked, and carrying what-ifs to their logical conclusions.

❖

Although SETI was safe, for the moment, the team still needed to be frugal—to use the small sum they had, while they still had it, as efficiently as possible.

"There has to be a way to stretch our funding and have room to build some real instruments," Billingham said.

Billingham and Barney Oliver—the Hewlett-Packard vice president who'd been involved with SETI since his meeting with Drake in Green Bank—began business talks, and Oliver agreed (tautologically) to fund a study about funding. He could bring in outside help, as his administrative assistant, Elyse Murray, happened to be dating (and later would marry) a man named Tom Pierson. Unlike the SETI team, who learned business practices ad hoc, Pierson had an actual business degree and fiduciary credentials. Through Murray, Oliver contracted Pierson to draw up a set of business models.

"Here's the money we have," Oliver told Pierson. "Here are its constraints. How can we do more with the little we've got?"

Pierson came back a few weeks later, after studying the situation, and told them one easy solution existed: create a new organization, just for SETI. All members of the SETI team were doing their work at NASA's Ames Research Center. The center paid for their office space, the light bulbs, the pencils. But only two of the SETI scientists were actual Ames employees. Everyone else, including Tarter, actually worked for one of the local universities, to which they had to pay part of their grant money, and just kept offices at Ames.

"Indirect costs," schools call this. Better known as "overhead," these charges added 80, 90, and sometimes 100 percent or more to the grant money the project required, and compensated the university for the infrastructure they provided to support the research. But that's not always where the money went. And the SETI researchers believed their funds could be better spent. Pierson suggested the SETI scientists form their own institution—the SETI Institute, a nonprofit, no boats. Their institution's overhead could reflect the actual cost of doing business, buying pencils, and keeping the lights on.

"Why don't you take a crack at writing the organization's charter?" Billingham asked Tarter.

She knew nothing about writing charters, but she generally believed she could do anything whether she knew how or not, whether that something was taking a radio apart or writing a grant. (This is a PhD phenomenon—because an academic knows everything about one thing, they tend to believe they are or could be expert in basically anything else.) So she sat down and started typing. "I wrote it as broadly as I could," she says. "We would be a research support home for scientists who wanted to tackle research having to anything to do with any factors of the Drake equation." Recall that the Drake equation estimates how many communicative, technological civilizations should exist in our galaxy, based on how often stars form, how often they form planets, how often those planets are habitable, how often those planets are inhabited, how often the inhabitants get smart, how often they decide they want to have interstellar conversations, and how long they survive as a technological civilization (whew). So SETI Institute scientists could study stars, planets, geology, biology, anthropology—anything, really, as long as they could connect it to a term in the equation.

That charter—which led to the current mission statement, "to explore, understand, and explain the origin and nature of life in the universe, and to apply the knowledge gained to inspire and guide present and future generations"—became the founding document

LEFT: Jill's parents, Betty and Dick. BELOW LEFT: Early on—there was balance: dresses and fish. BELOW RIGHT: Once Jill learned to braid her own hair, fishing, camping, hunting and building camp sites dominated. *All images courtesy of Jill Tarter.*

ABOVE AND BELOW: Learning to use the 300-foot telescope at the National Radio Astronomy Observatory, with Jeff Cuzzi (1976). *Both images courtesy of Jill Tarter.*

ABOVE: NASA SETI Science Working Group (1984). BELOW: Leading the Microwave Observing Project Team (1989): Dan Werthimer, Ed Olsen, John Dreher, Peter Backus, and Jill Tarter. *Both images courtesy of Jill Tarter.*

ABOVE: Third U.S.-U.S.S.R. SETI Decennial Meeting (1991). *Image courtesy of Seth Shostak.*
BELOW: First light at Parkes Observatory (1995). *Image courtesy of Jane Jordan.*

ABOVE: The Parkes Observatory. *Image courtesy of CSIRO/John Sarkissian.* BELOW LEFT: The Mopra radio telescope. *Image courtesy of CSIRO/ATNF.* BELOW RIGHT: The 140-foot dish at the National Radio Astronomy Observatory. *Image courtesy of NRAO/AUI/NSF.*

ABOVE LEFT: The dishes at the Georgia Tech's Woodbury Research facility. ABOVE RIGHT: The student-built feed at Woodbury. *Both images courtesy of SETI League.* BELOW: The Lovell Telescope at Jodrell Bank Observatory. *Image courtesy of Jodrell Bank.*

The Arecibo Observatory. *Image courtesy of NAIC/NSF.*

ABOVE: Astronomers don't get to experience this perspective. *Le Monde* had to take out a one-day, multi-million dollar insurance policy so that Jill could climb the towers with photographer Louis Psihoyos. BELOW: Inside the control room, the work was far more mundane—just baby-sitting the software intelligence with Jane Jordan to make sure it was doing the right thing. *Both images courtesy of Louie Psihoyos.*

Jill with Jodie Foster. *Image courtesy of Warner Bros. Francois Duhamel.*

The process of building a "Large Number of Small Dishes" (LNSD) array is a process of starting with commercially available materials and then optimizing. TOP: The Production Test Array. *Image courtesy of Seth Shostak*. CENTER: A dish is assembled from pieces in the construction tent. *Image courtesy of Dave DeBoer*. BELOW LEFT: A dish being attached to its pedestal in the field. *Image courtesy of Dave DeBoer*. BELOW RIGHT: *Image courtesy of SETI Institute REU Program*.

ABOVE AND BELOW: The Allen Telescope Array. *Both images courtesy of Seth Shostak.*

Four men who have supported SETI and Jill throughout the years: ABOVE LEFT, John Billingham. *Image courtesy of Seth Shostak.* ABOVE RIGHT: Barney Oliver. *Image courtesy of Hewlett-Packard.* BELOW LEFT: Frank Drake. *Image © Roger Ressmeyer, Getty images.* BELOW RIGHT: Jack Welch, *Image courtesy of Seth Shostak.*

A blended family. LEFT: Jill and Jack wed in 1980. BELOW: Back L-R: Jeanette Welch, Leslie Welch, Sandy Schniewind, Marni Welch, Shana Tarter. Front L-R: Mark Abbott, Jill, Jack, Eric Welch. *Both images courtesy of Jill Tarter.*

Eric and Marni Welch.
Courtesy of Mark Abbott.

Clara Welch. *Courtesy of Mark Abbott.*

Leslie Welch and Mark Abbott.
Courtesy of Mark Abbott.

Craig Kletzing and Jeanette Welch. *Courtesy of Mark Abbott.*

Steve Platz, Li Platz, and Shana Tarter. *Courtesy of Shana Tarter.*

Jill Tarter and Jack Welch. *Courtesy of Mark Abbott.*

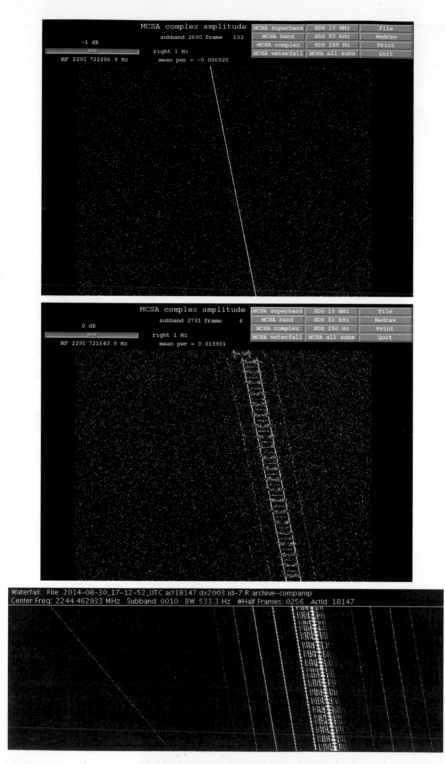

TOP AND CENTER: Carrier and sideband signals from Pioneer 10. *Both images courtesy of SETI Institute—Project Phoenix.* BOTTOM: RFI from SOHO.

Computers do the work, but hands can help explain it. LEFT: Jill at Arecibo with Kent Cullers. *Image courtesy of Jill Tarter.* CENTER: Jill at Parkes with Lee Hendricks. *Image courtesy of CSIRO/ATNF.* BOTTOM: Jill at ATA with Jack and Steve Trimberger. *Image courtesy of Barbara Vance.*

Working at the Phoenix Project.
Above left: Image courtesy of Seth Shostak.
Above right: Image courtesy of Ly Ly.
Center right: Image courtesy of Seth
Shostak. Below left: Image courtesy of
Jane Jordan. Below right: Image courtesy
of Jane Jordan.

Life after breast cancer. *Both images courtesy of Seth Shostak.*

Jack's retirement—everybody samba! *Image courtesy of Jill Tarter.*

Jill with her granddaughter, Li. *Image courtesy of Shana Tarter.*

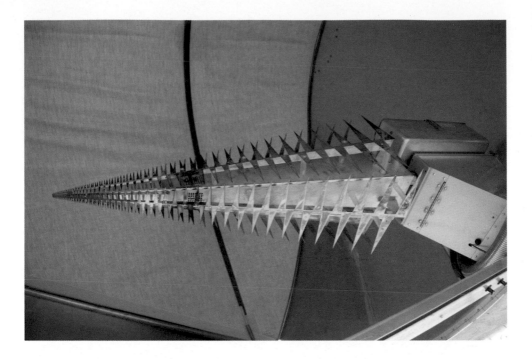

Original (above) and new cooled Antonio feeds (below) work inside the antennas.
Above: image courtesy of Seth Shostak. Below: image courtesy of Matt Fleming.

Jill retires at SETI Institute. *Image courtesy of Seth Shostak.*

Time to do something different, and time to continue what we started.
Both images courtesy of Jill Tarter.

of the SETI Institute, which still exists today and at which Tarter worked until her retirement in 2012. Although the scientists originally conceived the organization to save NASA money, its incorporation also allowed SETI as a science to survive the opposition that would come later. It separated them from the government—so they could fundraise and accept donations, which governmental research organizations can't—but allowed them to use federal funds, too, by applying for NASA research grants in subjects that were not explicitly related to SETI. As a separate entity, they could seek private money and do risky projects NASA and the National Science Foundation didn't approve of but which turned the heads of the public and a few key philanthropists.

"Did you feel like you were doing something important then?" I once asked her. "Did you have a sense it would save SETI?"

Tarter shook her head and repeated "No" a few times. "It was just the right thing to do at that moment to move forward, and that's generally what we did," she said. "We expected that a year or two in the future, we would have to invent a new right thing."

She paused, looking over at my notes. "I guess I had no vision," she said. "Period. I had no vision."

Pierson drove the SETI Institute's founding documents—Tarter's charter—to Sacramento, the state's capital, on November 20, 1984, in his old Honda. He worried the whole way. Could he find the right offices in the bureaucratic maze? Had they filled out all the right forms? Where would he type alterations if he needed to? But the SETI Institute was successfully founded at 2:00 P.M. that same day. Tarter was the first employee.

Tarter submitted the SETI Institute's first research proposal just in time for NASA's December 1984 deadline. The document created a unified team out of all the people who had been helping with SETI at Ames, under one institutional umbrella. A few months later, NASA gave the SETI Institute good news: the grant had been successful, and they could start work in February 1985. At a meeting later in Green Bank, a number of outside scientists opined about how only tenured professors who'd spent their earlier careers studying other topics should do SETI, rather than having a dedicated team at various career stages—that was the way to keep it safe and taken seriously. "Trying to raise funds for people to do this as a vocation as opposed to an avocation was not realistic," Tarter says, recalling their sentiment. And she worried for a while that she'd helped take her field on the wrong path, but later thought better of it and carried on.

And so never having been a team member, Tarter was suddenly a team leader: the head of the first SETI Institute project, which was known as Peterson's Left Leg.

Working above people is not necessarily her strongest suit. Although Tarter is generous and has a soft side, that side is sometimes hidden beneath something harder. Tarter is inspirational, says Gerry Harp, who took over as director of the Center for SETI Research when Tarter retired. But in day-to-day matters, her persona can grate. "She is not intentionally abrasive but she is just so focused on what we are doing and the SETI search that she isn't much of a touchy-feely kind of boss," he says.

But leading the charge, Tarter felt hopeful, like maybe SETI could finally have a firm ground to stand on, with its single limb.

❖

Peterson's Left Leg was the pet name for the old multichannel spectrum analyzer (MCSA), the prototype SETI instrument that split the radio signal into stations. But to find out what those cosmic radio stations were playing, the scientists needed to do a complex transformation of the radio waves. Called a fast Fourier transform (FFT), the operation changes data from a time series into a frequency series—in other words, it takes the signals that come in over a range of times and adds together all the signals with a frequency of 1,000 hertz, 1,001 hertz, and so on. It's kind of like making a bar graph that shows how much power is coming in at each frequency. On their own, Tarter's team had developed a system that could flip-flop 72,000 1-hertz-wide channels. But, as always, they wanted more. However, a chip did not yet exist that could do *fast* FFTs on big datasets. "Well, we'll just make one ourselves," Tarter said. Or, rather, they would ask engineers at Stanford to make one for them.

She knew that Stanford's Allen Peterson had Department of Defense money, and some of his graduate students, under astronomer Ivan Linscott, had been tooling around with FFT circuits. The engineers claimed they could create a chip for SETI that could process 10 million channels at once. This innovation had implications for industries like medical imaging (and, as always, defense), and so Linscott and graduate student Jay Duluk created a company called Silicon Engines to commercialize the technology, and electrical engineer and computer scientist Dave Messerschmitt helped oversee the collaboration. Despite their Silicon Valley roots, they ran into the same problems as most technological companies. The chip took longer than expected, and it went over budget. "Wringing of hands and clenching of teeth at NASA headquarters about not meeting milestones," Tarter says of the time, laughing.

NASA is obsessed with milestones, conceiving of deadlines as validations of moral character, despite hardly ever launching a mission on time or on budget. But Silicon Engines did eventually produce the chip, which measured 8 × 8 millimeters. But it housed 34,000 transistors and could do 80 million operations every second. That was revolutionary in 1987 (today, laptops can do billions).

"We're building a new machine to do some serious mining," Linscott told the *Stanford Daily* in a press release at the time. "It's virgin ground, and who knows what we'll find, but I think we have a pretty good chance of detecting extraterrestrial signals."

(Optimism, and probably stubbornness, kind of by definition, walk hand-in-hand with SETI. As Bill Borucki said of Tarter, "Someone who keeps going even though things look very bleak sometimes: that's what you need.")

When they did actually reach their milestone, Lynn Griffiths, the NASA headquarters program officer for SETI, called a meeting to say there were no hard feelings about how they hadn't done it *on time*. "She came out from headquarters intending to smooth things over and say, 'Atta boy,'" says Tarter. "She even had a rock in her purse on which she'd written something to commemorate their achievement, their mile*stone*."

But in true Silicon-Valley spirit, Linscott and Duluk took this milestone meeting to be a product release, and wrote a press release to accompany it, not mentioning NASA or the SETI Institute. The recriminations that followed, and further missed milestones, tore the start-up apart, with the partners bickering about who would buy out whom. The tattered remnants of Silicon Engines manufactured the crucial chips for the SETI Institute and then dissolved completely.

❖

It was a work hard, play hard time in Tarter's life. After each summer American Astronomical Society meeting in this era, Tarter, Welch, and a few scientific friends would charter a boat in the British Virgin Islands, where they would sail for a week. They docked only to eat dinners, and sometimes not even then. On one trip in 1988, astronomer Margaret Burbidge joined. She was an astronomy pioneer before Tarter was conceived—the director of the Royal Greenwich Observatory, the first woman president of the American Astronomical Society, president of the American Association for the Advancement of Science. She had gained access to California's Mount Wilson Observatory in 1955, when it was still reserved for men, by working as her husband's night assistant. In so doing, she opened the mountaintops to all women.

Tarter often sat on the edge of their boat during the day, embroidering their course onto a touristy scarf on which a map of the British Virgin Islands had been printed. As she stitched the boat's path she joked with Burbidge, one of her personal heroes. They sailed for only a few hours every afternoon, and then searched for fresh fish. And as the sun dimmed, the rum rose. They jumped into debates about topics like the origin of Shakespeare's plays and deep observations of pelican behavior.

In pictures from the trip, the water seems to emit its own blue light, rendering the boat's passengers otherworldly in their happiness. They are young. They are a little drunk. They are standing on the hull of a boat. They imbibe and argue long into every night, with Earth's particular view of the sky above, the oscillation of its oceans below. They do not care, at this moment, that one day they will grow old, that these sailing trips will end and become something they only look back on. They seem to live just in this one moment of Earth's history, occupying this particular layer of the rock record and not thinking about the ones below, or the ones that will one day press down from above.

Tarter once said that she has mostly lived her life thinking that she will be immortal, but the specifics of aging do worry her. Whenever she forgets where her keys are, she wonders if it's a normal lapse or the beginning of something sinister. Every time she forgets a name, she mutters, "Damn anomia," and then, as a directive, "Never get old." One of her favorite sculptures is Rodin's *She Who Was the Helmet-Maker's Once-Beautiful Wife*.

"This old, old, old woman," she says, "but you can still see that she was once quite beautiful."

❖

By the time her mother was doing SETI science under the SETI Institute, Shana was in college at Cornell, her mother's alma mater, with a year abroad in Durham. She brought a plastic pink palm tree with her, to remind her of California and the Pacific during Ithaca's long winters and Britain's rainy springs. During the summers before and after her academic year, she worked in remote southwest Scotland, excavating the first Christian church there. She and her pretty, long-haired boyfriend, Martin, lived in an old, damp cottage. There was no plumbing, and electricity only if they put coins into a box outside. The couple dug for evidence of ancient intelligent humans during the day and frequented the local pub in the evenings, where they also paid in coins for weekly showers. Tarter and Welch came to visit both summers, scooping Shana and Martin up in their rented Saab and whirling them around the island.

When Shana returned to Ithaca and graduated from college, she spun a cartwheel, just like her earliest (and Tarter's last) school days. Tarter's mother, Betty, sat in the audience, happy that at least one of her descendants would consent to a ceremony. "I never went to any of my graduations," Tarter says. "I was always on to something else."

At graduation the whole family coexisted for a few days in and around Shana's tiny house in Varna, just northwest of Ithaca, to celebrate the occasion. Welch and Bruce, perhaps in an effort to deal with each other, were on a search for alcohol. No one would sell them any, since it was Sunday, and that was the law. Eventually, Shana reached under her kitchen counter and pulled out a bottle of tequila, its worm tossed by the currents in the viscous liquid, like an alien in a jar of mercury.

Bruce took it from her and poured them all a shot.

To growing up, to intelligent lives.

❖

But the circle of life happens to everyone. One day in 1988, while Tarter was away using a telescope in Nançay, France, the phone rang in her hotel room. Welch's mother had died in her sleep.

Tarter had never been to her husband's hometown, but along with all Welch's children they both traveled to Westfield, New York, to commemorate Ruth Welch's 94 years on this planet. Tarter had never met the other celebrants, except for Welch's sister Judy, but from the long nights she'd spent drinking scotch liqueur with Ruth, she knew that the ceremony would have pleased the rule-breaking operatic star.

Jack Welch's family is, in fact, the Welch family of the grape juice. The company began with his great-grandfather, a dentist and also a

Methodist. Methodists in that era did not believe in wine, or at least didn't believe in drinking it, so they used grape juice for communion. But unpasteurized juice doesn't stay potable for long. One day just after the Civil War, the dentist read a paper by a man named Louis Pasteur. The elder Welch began experimenting with Pasteur's germ-killing methods, boiling grape juice at home and bringing it to church to share. Soon, Welch's Grape Juice became the second pasteurized product (following condensed milk) to appear commercially.

Welch's father, however, wanted nothing to do with this juice or the family business. Instead, he took over a flexible coupling company in town and then became a deputy sheriff during Prohibition—even though (or perhaps because) Canadian bootleggers used the Welch's Grape Juice lighthouse to get across Lake Erie. Someone later shot him in the ass while he was on duty, but he survived.

Nonetheless, because of this beverage legacy, Tarter and Welch inherited a large sum of money when Ruth died. They decided to use it to buy a house at Donner Lake. Together, they have used some of their extra money—like from Welch's consulting work on the Allen Telescope Array and as part of Tarter's role as a trustee—to help fund SETI, donating around $300,000 to the institute.

And the Donner house, which they still own, is now surrounded by the newly built mansions of retired Silicon Valley millionaires (and perhaps billionaires), much fancier than their own building. In 2015, a 130-mph windstorm snapped the top off the lodgepole pine that rises through their house's three stories of decks. With the remaining trunk, Tarter plans to create a "Tarter-Welch totem." A man skilled with a chainsaw will fly in from Boulder to sculpt different animals on each level—bear, osprey, owl (the familial icons for Welch and Tarter), and other local fauna—stacked in vertical layers like the fossils.

CHAPTER 7

THE QUEST FOR CONTACT

T arter settled onto the speaker's stool in the Burbank sound studio. "On the surface of an average planet," she began reading, "circling an ordinary young star, an advanced intelligence searches the skies for evidence of life."

Director Geoffrey Haines-Stiles's voice appeared in her ear, and his arm waved like a windshield wiper across the glass in front of her. He had written the script for the 1988 movie *Quest for Contact*, a romp through the cosmic haystack with narrator Jill Tarter. And she was butchering it.

"No, no," he said. "Like this." He repeated the sentence back to her, the way she believed she'd just said it.

She paused and spoke back into the microphone. "On the surface of an average planet, circling a—"

"Cut."

This happened over and over. The director's intonations sounded the same to her.

Years later, when Jodie Foster narrated an education video for the SETI Institute, she showed up dressed in sweatpants and covered in baby spit-up, freshly off a plane from France. But Foster breezed into the studio, put on earphones, and read the script once without pause. She shook her hair out and read it again. And then she walked out of the booth and handed the script back.

"Anybody who knows what they're doing can get whatever they want from these two readings," Foster said.

The two takes sounded the same to Tarter, who can't tell that a middle C is lower than a middle G, or the difference between mā and mà in Chinese. "I haven't ever developed a voice," she says.

But as the head of NASA's new SETI program, Tarter would have to. In 1988, Congress authorized the SETI Institute's Microwave Observing Project (MOP) to receive $6.6 million of NASA's budget for the 1989 fiscal year. After Congress agrees to dispense money, though, an "appropriations committee" actually decides how to pass that money out. It's like asking the good-cop parent for an allowance—they say yes, but when bad-cop parent finds out, they may rescind the offer. That year, the appropriations committee decreased MOP's funding by two thirds, to $2.2 million. It wasn't enough to build any physical equipment that could actually start the search for extraterrestrial intelligence.

It makes sense that SETI isn't politically popular. It doesn't create many jobs; it doesn't feed starving children. Many people, including politicians, thought of SETI as useless at best and science fiction at worst. Tarter wanted that to change. And to make that happen, she had to go to DC.

❖

On lobbying trips throughout 1989, Tarter learned well the monochrome hallways and mahogany desks of the nation's capital. The low-level staffers, often sci-fi fans, helped Tarter find the tools to navigate Capitol Hill. Most key: books that listed committees, other staffers, politicians' specialties, and room and phone numbers for senators. The staffers slipped Tarter information about secret hot-topic issues and insider knowledge of whose son-in-law worked in which state's aerospace company.

Tarter marched from appointment to appointment, riding the waves of support and dismissal. Some senators thought SETI was a fine idea. But others said it was irrelevant. Aliens, if they exist at all, live far away. And SETI can't ever—in a million years—guarantee a slam-dunk, let alone guarantee one in a senator's lifespan, let alone in a politically relevant timeframe. SETI may be philosophically appealing, but contemplating philosophy is one thing and investing constituents' tax dollars in it is quite another.

As Tarter walked out of each congressional office, feeling cos-mically diminished, she often saw a silkscreened Gravity Probe B T-shirt flung over admins' chairs. It was swag left by Francis Everitt, who had passed through lobbying before her. Scientists had first proposed the Gravity Probe in 1959, to measure and test Einstein's general relativity. Development had spread and sputtered over decades. The project's lead scientist, pleading for funding and mercy, was apparently trolling the same offices as she was, a few steps ahead, trying to ensure that his 30-year-old dream didn't die. And it worked; Gravity Probe B launched in 2004, operated on orbit for 18 months, and delivered its final validations of Einstein's theory of general relativity in a special issue of *Classical and Quantum Gravity Journal* in November 2015.

The planning stages of "big science" projects, which can require teams the size of towns, often span decades. A scientist can spend their whole career developing an instrument that NASA, the National Science Foundation, or the National Institutes of Health cancels at the last minute. For scientists' own sanity, they have to be sure, upon setting out on their career paths as bright-eyed twenty-somethings, that when they look back on their lives, they will be happy with how they spent their days—even if they spend most of those days on a project that gets canceled. In SETI's case, even if Congress builds telescopes and those telescopes work perfectly, a null result—no beacon from another civilization—remains a possibility. Tarter has taken joy in the search, maybe even in its drama. And when she began, she and most other scientists, from Drake to Billingham and Oliver, felt they might succeed quickly. But after a few years in the field, she knew the *Encyclopedia Galactica* might never appear in their telescopes—and that they might not even ever get to build those telescopes and try.

Congress restored SETI's funding to $4 million per year late in 1989. Tarter spent so many nights at the DC Holiday Inn that they started to send her coffee mugs each Christmas, and she amassed enough frequent-flyer miles that TWA gave her an around-the-world trip—Hawaii, India, Paris—for free, for both her and Welch.

It once puzzled me to think that Tarter had gone to grad school in Berkeley in 1968, bought a house there in 1980, and never left. It seemed provincial, like the hippie version of living your whole life in a tiny hometown. Why hadn't she wanted to live other places, meet different people, and build decks on houses in other states? And then she handed me a printer-paper box containing decades of calendars. A full 30 years of her life look like a constant around-the-world trip. For her, it seems home was the novelty.

❖

Upon their return to the United States, Tarter leapt back into the new SETI program. "We needed to prove to the astronomical community that we knew what we wanted," she says. "We put the NASA proposal together over and over and over again. We made pre-project plans, and we made project prototype production plans. We wrote a plan, and then the rules of the game changed, and we'd write another plan. And we'd write another plan."

And Barney Oliver was always there to correct their grammar. His mother had been a schoolteacher, and Oliver was known for derailing public speakers with such semantics. As a lecturer stepped merrily along the sentences of their presentation, they inevitably came to a line like "I'm anxious to see the results." Not a nanosecond later, Oliver's voice would boom from the back: "Eager," he would shout.

The speaker would look around, confused. "You're eager," Oliver would clarify in no softer a voice, "not anxious."

He took the same linguistic stick to SETI's pre-project, project, and prototype plan write-ups. "I had a tendency to say *which* when Barney said I should be saying *that*," Tarter says. "I'd have an assignment, I'd write it, and then I'd sit down with Barney and he'd change all my *which*es to *that*s." They both called these syntactical excursions "which hunts."

But one day, Oliver tried to change her *anticipate*s to *expect*s. "You can't do that," she told him, shaking her head. "I'm female, and I'm very devious. I'm going to do whatever is necessary, behind the scenes and behind your back, to make sure it happens. So I'm not expecting that outcome, goddammit, I'm anticipating it." Grammar be damned. They had to get those grants.

Around this time, in 1987, Tarter also discovered that someone with the same name as her undergraduate lab partner—Leonard Fisk—had become the associate administrator of space sciences at NASA. Tarter ordered an old yearbook to see if this was the same Len. It was. It felt good to have an ally.

❖

The NASA-funded SETI Institute team began to work on two separate, complementary searches for extraterrestrial intelligence. In a November 1989 memo, the NASA SETI program manager, Lynn Griffiths, said of this full-on strategy, "If we have to go down, let's let 'em know they've been in a fight. Damn the torpedoes. Full steam ahead."

Tarter would head up a "targeted" search, to deeply scan sun-like star systems within 100 light-years of Earth. By focusing on these near neighbors, the team could look more closely at potential civilizations whose broadcasts might still being strong enough when they reached Earth (just as your favorite station fades when you drive too far, extraterrestrial missives get weaker the farther you are from their origin). Their search, at frequencies from 1 to 3 gigahertz, would cover trillions of times as much ground, data-wise, as all of the previous SETI experiments combined.

Bruce Murray, the director of the Jet Propulsion Laboratory a few hundred miles south in Pasadena, would lead a "survey" that looked more shallowly over the whole sky, from 1 to 10 gigahertz. This survey aimed to turn up the behemoths of alien broadcasters, whose messages were sent so strongly that we could see them from far, far away.

The sky survey equipment was being developed at JPL, while Stanford and the SETI Institute were developing the targeted search equipment. Tests happened at NASA's Deep Space Network Goldstone site. So every Tuesday, Tarter and her colleague Peter Backus boarded a NASA 7 aircraft at Ames. Onto this plane, Tarter loaded three days of work clothes—T-shirts with science jokes or space mission logos, jeans, Asics—and a pile of journal articles to read on the hour-long flight. Once they landed at Dryden Flight Research Center at Edwards Air Force Base, she and Backus leapt from the roll-up staircase and ran to the motor pool to pick up a car for the two-hour drive through the desert to the Goldstone complex at Fort Irwin. There, they dumped their gear in an old converted barracks and headed for the telescope and lab where Peterson's Left Leg lived. They fiddled around with code and receivers and reseated circuit boards, preparing the equipment that would prove they could actually do what their years of drawn-up plans proposed. Once, they discovered a mouse harvesting cable insulation for its nest. After that, the team chipped in to help the station operators house and feed an unofficial—and undisclosed—cat.

Fifty-six hours later, on Thursday afternoon, Tarter and Backus navigated the desert back to Edwards, hopped back aboard the plane, put their figurative seat backs and tray tables in the upright and locked positions, and curled against the windows to sleep for an hour. It was like being the child of divorced parents: always schlepping between home bases, beset by two sets of expectations, wondering occasionally whether it wouldn't be better to just be a shopkeeper.

❖

One day, Tarter, Backus, and the JPL team finally got the spectrometer and the rest of Peterson's Left Leg working. It could scan the sky for radio signals, split them up into tiny frequency bands, and show how they changed over time—or so the team hoped. To demonstrate it, they used their usual strategy: test the instrument's ability to find intelligent extraterrestrials by seeing if it could detect intelligent human technology. This time, a spacecraft.

The *Pioneer 10* probe was, at the time, the farthest-away hunk of human-made metal. It had traveled billions of miles from Earth, beginning in 1972, and revealed the details of our own solar system for the first time as it flew by planets no other human-made object had ever been close to. Jupiter had a stormy blemish, *Pioneer*'s images showed; Saturn's rings were full of holes; Neptune was nothing but blue. And then *Pioneer* continued onward, into the great black beyond us. Its far-off signal was weak. The SETI equipment would only be able to detect the spacecraft if everything functioned perfectly.

On the day of the demo, Oliver arrived just before their first attempt to find *Pioneer*. Prepared for the worst, he stood at the periphery of the group, chewing on his glasses. He clenched his jaw, his teeth grinding against his glasses. All of the scientists watched him and the screens, waiting. They could feel each other breathing.

And then, on the screen, a little white diagonal line sliced across the static. The white, standing out against the gray static, meant the signal was louder than the random noise around it. Its shape—a diagonal line—is just a reflection of the Earth's rotation and of how the software displays data. The vertical axis shows the progression of time; the horizontal axis shows the radio waves' frequencies. So a white pixel halfway across the screen and midway down says to scientists that something emitted a lot of radio waves in the middle of the frequency band, in the middle of the observation. A white line down the screen means something sent out radio waves at one frequency for the whole time; a white line across the screen means something sent a signal that came at all frequencies at just one moment in time.

If the detector were right next to *Pioneer*, that former plot represents what it would see from the spacecraft: A white up-and-down slash—a frequency-specific "drone" that is continuous. But Earth's rotation skews the frequencies that arrive on Earth from space. So even though *Pioneer* sends radio waves at just one frequency, it looks, from Earth, like that frequency shifts a little bit every second. *Pioneer*'s valiant pings—and the pings and blasts of anything outside our atmosphere, be they supernovae, satellites, or civilizations—show up as narrow arcs or wide tracks down and across the screen.

That's all a "signal" of any kind is: A white slash across a grayscale background. That was *Pioneer*. And that is how a communication from extraterrestrials—one of the most profound discoveries in human history—would look. Very simple, in other words. If there is a message encoded in the signal, it will be spread across many frequency channels and thus much harder to find, or perhaps embedded in the way the signal's waves are oriented. Scientists might spend years, or whole careers, extracting the information required to turn that slash into something sensible, if it came from actual aliens.

Astronomers have long sought to *create* messages that extraterrestrials could decode—simple, scientific visuals like those in the radio communique Frank Drake and his colleagues sent from the Arecibo telescope in 1974. Embedded in a radar broadcast was a counting lesson in the language of octal, symbolizing the numbers 1 to 10. Following that, using the same symbols, Drake placed the atomic numbers of the elements that make up the components DNA; below that and using *those* symbols, he broadcasted the chemical formulas for the more complex sugars and bases that make up the nucleotides in DNA; below that and using *those* symbols, Drake finally got to the meat of the matter: the number of nucleotides DNA has and a double helix that was almost comical in its old video gaminess. A human figure from the same pixelated video game followed, as did a schematic of our solar system and a representation of the radio telescope itself, complete with a boast about its diameter.

The message attempts to teach its recipients how to read it. But, as the SETI Institute's former director of interstellar message composition, Douglas Vakoch, often says, these beings might not have symbols in their culture—abstractions, like pixelated shapes, that stand for something concrete, like numbers. Hell, they might not have numbers. Double hell, they might not have *eyes*. Or any of our other senses.

Those tasked with composing interstellar messages—or at least thinking about composing them—like to cite philosopher Thomas Nagel's 1974 essay "What Is It Like to Be a Bat?" whose central thesis is "You will never know." You can imagine what using echolocation feels like, think about hanging upside-down all day, but even if you mastered those feats, they are just bat behaviors. You can never really— *really*—understand a bat's experience of the world. And bats, while not strictly "like us," at least evolved on our planet. Extraterrestrials could be more different from us than the very strangest organism on Earth.

But, back at Goldstone, the team knew what the white-slash signal meant, because organisms very much like them—NASA engineers— had made it. It communicated, simply, "I'm here; you can see me." And they could. The scheme and the instruments had worked.

"It's an exciting time for us," Tarter told the *New York Times*. And speaking about the final, full project, which they hoped to launch in 1992, Backus told the same reporter, "In the first minute, we'll accomplish more than all the other projects combined."

From then on, the SETI team hunted down *Pioneer 10* every morning before starting work, as a test. If they could find humans, the thinking went, they might be able to find Others.

❖

But next, they had to test their alien-hunting equipment on the actual telescopes where they would actually be doing the search. So Tarter, Backus, and Larry Webster, the SETI program manager at Ames, trudged through the San Francisco airport. The big steel suitcases full of SETI instruments strained their elbows and made their torsos list to one side. They were glad to check them and send them off with a big SYD sticker on them.

Their destination: Tidbinbilla, 205 miles from Sydney, a town that contains nothing much except the Canberra telescopes, which NASA built as part of its Deep Space Communications Network. The area lies in the "Australian Alps" (according to Australians) and is home to a nature preserve with animals Westerners consider exotic, like wallabies and kangaroos, the white-tailed deer of the Outback. Cows often meander around the telescopes, not caring at all whether outer-space signals slam into the dish above them.

Once they arrived in the telescope's control room, Tarter and Backus set their suitcase SETI equipment up on the desk. They felt smug and self-satisfied. Look at us! Such small equipment to do humanity's biggest project!

The hubris lasted only as long as it took to press the power button. Circuits sparked. Something popped. Smoke appeared.

They had forgotten what everyone who travels internationally with a hair dryer knows: The wall outlet dispensed 50 cycles per second/230 volts, not the 60/110 of the United States. They should have thrown the computer's trip-abroad toggle, a switch that worked as an adapter. But in their excitement and REM-bereft state, they had neglected that simple switch.

Webster tightened his facial muscles. He set his palms against the desk surface for a second. Then, silent, he lifted the case by its handle and carried it back to his hotel room like the deadweight it now was.

The next day, an Australian technician discovered that a capacitor had blown up early in the electrical path, and the deadly voltage never made it past that capacitor and into the rest of the electronics.

All they had to do was swap in a new capacitor and move on. Which they did, to the Parkes telescope. Miles and miles and miles they drove, down a narrow road with only Outback around them. Up ahead, though, Tarter could see something in the road. As they got closer, it resolved into branches—an outsized Australian tumble-weed, taller than the car. The driver stopped, and Backus ran out and pushed it out of the way, like a boulder.

❖

Although the SETI program made sojourns abroad and moved toward maturity, members of the US House continued to speak out against MOP. They compared SETI scientists to the UFO enthusiasts and the unfortunates who believed their eggs had been harvested by aliens. In one case, a congressman from Tennessee put on a pig's nose and talked about a pork-barrel project—referring to NASA generally, SETI specifically. In June 1990, Ronald Machtley of Rhode Island and Silvio Conte of Massachusetts took official action, introducing a motion to cancel all SETI programs the next fiscal year. Conte asked the somewhat chilling question, "Can we afford curiosity?"

Senator Barbara Mikulski, the appropriation committee's chair-person, believed they could. "The committee reaffirms its support," she said, "of the basic scientific merit of this experiment."

But in May 1991, Senator Richard Bryan of Nevada aimed to stop SETI in the 1992 fiscal year, with an amendment to cut it from the authorization bill. He called it "the $14.5 million Martian hunt." In a May 1991 press release, he elaborated: "At a time when our country faces massive budget deficits, urgent health care needs, and inadequate educational funding, federal government has no business financing something as superfluous as this." Funding for fiscal year 1992 was restored in a joint conference committee, but when Congress began to consider the '93 budget, the wrangling started all over again.

Tarter braced herself for another game-over.

But a senator from Utah—a Mormon named Jake Garn—was now on their side. Mormons—members of the Church of Jesus Christ of Latter-Day Saints—believe that if they behave themselves in "this life," they will be given a celestial reward in "the next life": each faithful congregant will inherit their own planet. They will be the gods of this planet, which will have its own population. Mormons have long believed in an abundance of exoplanets, populated by smart, religious beings. "As man is, God once was," the doctrine goes. "As God is, man may yet become."

Mormons would be happy to get in touch with some of those other worlds, and prove themselves right. Senator Garn served as ranking minority member of the Senate Appropriations Subcommittee, and they formed a formidable partnership to obstruct Bryan.

A Mormon-SETI alliance, or any collaboration between those who require evidence and those who don't, seems strange. But on the door to Tarter's current office hangs an inspirational decoration that Garn might have liked. A set of faux-aged boards, strung together with rope, displays the words

No whining

Be happy

Do your own thing

Enjoy life

Believe

Dream

Big

One difference, though: Next to *Believe*, however, Tarter has written "(but verify!)" in black Sharpie.

❖

In April 1992, Congressman John Duncan of Tennessee introduced an amendment to strike down SETI's 1993 plans. He had read an AP article about "setting up some SETI equipment in the Mojave desert to look for space aliens," and he wasn't happy. The onslaught continued: In June, Richard Bryan brought an amendment to the authorization committee to terminate the SETI project. He wrote:

> *There are those in this town who say that $13 million is not a lot of money, but that shows how out of touch the process is. The $13.5 million that we save under this amendment is the equivalent of providing 10,135 students with full-tuition scholarships to University of Nevada-Las Vegas, buying 115 new homes in Las Vegas, or providing day care for 3750 toddlers.*
>
> *In an ideal world with unlimited resources, this program might be worth considering. I am a strong supporter of NASA and scientific research. However . . . we cannot afford a program as remote and uncertain as this.*

The Joint House/Senate Conference Committee would decide who was right.

❖

That committee, which included Mikulski and Garn, met for what seemed like eons. Finally, Mikulski's staff director for the appropriations subcommittee, Kevin Kelly, emerged with the list of approved projects. Everyone stood around like high school theater kids waiting for the cast list to be posted. People mobbed Kelly, as he stood on a dais to pin a paper bulletin above the bustle. And over the ocean of concerned citizens, he hooked Tarter's eye. "Thirteen point seven million," he mouthed, "with language."

This meant that not only did they get to keep searching, but their money couldn't be taken away and used on something else, like university donors can specify that their donations can only be used to plant new grass. The churning crowd blurred into a background.

"With language," Tarter mouthed to herself, a verbal obfuscation that she had never been more grateful to hear.

The appropriations committee told NASA to remove the SETI Project from the Life Sciences Directorate and to reconstitute it within the Space Sciences Directorate. And then to rename it—change MOP to . . . something that sounded even less like aliens than "MOP." So they pulled out their mirrors and blew some smoke: MOP became the even more innocuous High-Resolution Microwave Survey (HRMS), aliens hiding behind the acronym.

"He Really Means SETI," the team wryly joked about the acronym's origin. Or, depending on Tarter's mood, "Her Royal Majesty's SETI."

❖

MOP/HRMS was set to begin soon after that. They had made their equipment mobile: it would all live in a tractor trailer. They planned to haul that portable house, and the signal-detection technology it contained, to the telescopes they would use. They had customized the structure so that a military C-141 plane could airlift it from site to site, starting with the initial observations in Arecibo.

As the HRMS team prepared to actually start searching, the *LA Times* asked Tarter if she thought they would find anything. She spoke about the SETI Institute's technical prowess, highlighting hertzes and algorithms.

"I expect to be successful at that search," Tarter told the reporter. "Whether that is sufficient to produce a signal, that is another question. I think anybody who is working on this project has a very good concept of just how enormous, how vast, this search that we're starting really is. We hope to be successful in our lifetimes but understand that maybe it's going to be our children that succeed in detecting the signal."

Many years later, while she and I sat on the floor of the SETI Institute going through decades of archival files, she added a generation to that same sentiment. "I can't guarantee that I'll be around for the end of SETI—only our grandchildren may be," she said. "But I can tell you that being around for the start has been fun and educational in ways I could never have predicted."

When she did that *LA Times* interview, the SETI team had been taxiing on the runway for more than a decade. They were finally ready for their flight to begin. To actually *do* SETI. They planned the HRMS opening ceremonies for Columbus Day, 1992—the 500th anniversary of that man's "discovery" of North America. The choice of date feels conspicuous and self-aware: made for press coverage. It would have been more prudent to keep the project's launch private and low-key, if they wanted to evade public and political ridicule. But Tarter objected. She refused to be pushed into a closet just because people disapproved.

"Hell, everything else at NASA has a celebration," she said. "We should have a launch party." Higher-ups like exobiology program manager John Rummel—who, along with Lynn Harper, had helped her run interference with NASA headquarters—advised her to keep it under wraps and under the radar.

"I didn't listen," she says.

❖

On Columbus Day Eve, 1992, Tarter paced the Arecibo Observatory control room, making sure every winding blue cable was in place, every signal pathway was sound, and every cryogenic dewar did its job. She looked out the panoramic window into the ancient sinkhole below. The giant radio dish—a mirror of mesh 1,000 feet across—filled the space perfectly. Engineers had picked the telescope's location by spreading out a topographic map of Puerto Rico and sliding a quarter around to see which valley could hold it. The quarter nestled precisely within a sinkhole 10 miles from the town of Arecibo. Three concrete pillars, which summer interns (and Tarter) occasionally climb to impress each other, rise from the edges of the basin, which the dish fills almost completely. Steel cables as thick as your forearm reach from the pillars toward the middle of the dish (although 500 feet above it). They hold aloft the radio-wave detectors and the electronics that make this huge contraption more than just a big bowl of chicken wire.

The Columbus Day Eve sun began to set, and the sky streaked the colors of an airbrushed '80s T-shirt. Maybe somewhere else, on some other planet, some other sky was streaked the same colors. Maybe someone was there to watch. These *someones* wouldn't know what the '80s or T-shirts were, but they would know starsets.

Tarter turned from the window and prepared to test the equipment with Backus.

"Ready?" he asked.

She nodded.

They pointed the telescope toward *Pioneer 10*, testing just like always. It showed up, a slash on the screen, just like always. Then, they turned toward a few stars—more tests. The computers they had built talked back to them, delivering good and unexpected news: they had found an interesting signal, interesting enough to "send a shiver of excitement through everyone in the control room," Backus told the *New York Times*. "Then it struck me," Backus continued. "Maybe what we were seeing on the screen is exactly what we are looking for. Sometime in the next couple of weeks we might do it for real. Who knows?" The signal turned out to be from a physical, not a biological, source

Tarter stayed in the control room until 3 A.M. When she walked back to her two-room hut, with its floral-upholstered couch and bamboo table, the chirping of the jungle frogs was deafening. But the natural noise was a welcome change, taking her mind for a moment off the nervous hum of electronics.

❖

A few hours later Tarter awoke and got dressed for the press. It was the day Her Majesty's Royal SETI began. She prepared to keep the *Cyclops Report*'s promise.

Outside the control room, where coder Jane Jordan's software prepared to search for alien signals, a crowd gathered, including Shana and her brand-new husband, who also took their honeymoon photos on the telescope's catwalk during the same trip.

Billingham stepped before the crowd to give an opening speech. He had spent even longer than Tarter waiting for this moment. "This is the beginning of the next age of discovery," he said. "We sail into the future, just as Columbus did on this day five hundred years ago. We accept the challenge of searching for a new world."

The audience, including the scientists who had worked for more than a decade to make sure someone like Billingham could make a speech something like this, smiled taut smiles and looked out toward the radio dish.

"If you're going to do this," Oliver had long ago told Tarter, setting a gold statue of Sisyphus and his boulder on her desk, "you're going to need this. Because you're going to roll an awful lot of rocks up an awful lot of hills, and they're all going to come tumbling down. And you're going to have to do it again. That's just the price of trying to do something new."

Tarter thought maybe the boulder had finally crested—today, Columbus Day, 1992. She pressed the buttons that told the telescope to start observing. "We begin the search," she declared. Simultaneously, Sam Gulkis did the same at the Goldstone telescope in California, starting the survey portion of the search.

The Arecibo Radio Telescope pointed at the star GL615.1A, 63 light-years away in the constellation Hercules. GL615.1A is like our sun but smaller and cooler. *God, this is a really amazing day for humans,* Tarter thought. *Here we are launching this exploration simply because we're curious. That's a big milestone for humanity. We're doing this.*

A *New York Times* reporter covering the event waxed philosophical, too, about the telescope itself: "There was speculation as to what future archeologists might surmise if they happened on the ruins of these stone pillars, aluminum panels and huge steel cables and girders. Here a society with scientist-priests communicated with their gods in the heavens?

Some Columbuses sought the cosmic Indies, never found? Or this was the place where humans listened in the jungle stillness and for the first time heard that they are not alone in the universe?"

❖

Senator Bryan, perhaps via this very *New York Times* piece (newspapers were always causing trouble for SETI), caught wind of the celebration. He had wanted SETI gone, and here SETI was, starting up in earnest. At a hearing for fiscal year 1994, Bryan's words sent a shiver through Tarter when she watched on C-SPAN: "Mr. Goldin," Bryan said to Daniel Goldin, the head of NASA, "something in your budget doesn't pass the smell test."

"He was talking about SETI," Tarter says.

Goldin says he was caught off-guard by the congressional opposition, in general, to SETI. As a new administrator, he knew the research program existed, but he didn't know much about its specifics. He says he wished someone had warned him about what he was walking into. "I was so frustrated that I had only a layman's understanding of the program," he says, "and I'm a detail person, and I always do homework before I do anything, and especially before hearings."

During that hearing, Tarter leaned toward the television, like it was a black box that could tell her future. Having knocked on as many White-House doors as she could, all she could do was wait for the final hearing, where people she didn't know would decide whether her career lived or died.

"It's hard to elevate the consciousness of Congressmen from mundane to heavenly matters," Barney Oliver once said in an interview with the *Times*.

❖

In September 1993, Congress met to talk about science and technology projects. To build solid rocket motors or to not build solid rocket motors? To build the superconducting supercollider (yes, a real thing) or to not build the superconducting supercollider? They had been going at it for days, slashing this and cutting that. Tarter watched C-SPAN for hours, thinking how much *more* boring it must be in that room. She switched off the television and went to pack her suitcase. She was scheduled to give a talk in Huntsville, Alabama, as part of the Wernher Von Braun Lecture Series at NASA's Marshall Space Flight Center. The whole night—meant for the public—was about exploration and the human spirit. Tarter would speak about SETI, of course, and folk musician John Denver would serenade the audience with world-uniting songs.

John Denver was a pilot, like Tarter. He loved to look down on Earth from a height where the horizon stretched long and curved, and countries were just pieces of connected land. His 1986 album was, in fact, titled *One World*. "My music and all my work stem from the conviction that people everywhere are intrinsically the same," he said in the folio *John Denver: A Legacy of Song*. "When I write a song, I want to take the personal experience or observation that inspired it and express it in as universal a way as possible. I'm a global citizen."

It's a hippie-era perspective, but one that space enthusiasts like Tarter also often embrace. That's part of why Denver was such a NASA groupie. And for his efforts in sparking interest in space projects, Denver was awarded a NASA Public Service Medal in 1985. After the *Challenger* disaster in 1988, in which schoolteacher Christa McAuliffe and six others died in a fiery explosion, he even dedicated the song "Flying for Me" to all astronauts, everywhere, always.

And as much as Denver liked NASA, Tarter liked Denver. "This cements me as an old hippie," she laughs, a statement she often makes about herself. "Like John, I'm a one-Earth protagonist."

She stood backstage as Denver performed "White Horses," swaying and watching the crowd do the same. They were all there together, in this moment in the dark in Huntsville, thinking about the long future, the big space, and their place in it all. It was kind of beautiful.

But at the same time, Congress sat behind long desks discussing whether to interrupt that line of questioning. "It was all overwhelming," she says in 2015, looking toward the wall of her Berkeley home, where the plaque commemorating the Von Braun lecture hangs. "I was overwhelmed by the star power on the stage and the DC shenanigans threatening to terminate my world."

Just before she was to succeed Denver on the stage, a staffer whispered in her ear: Senator Bryan had put in an eleventh-hour proposal to cancel the SETI program. Congress would vote in the morning. She calls Denver's performance a Rocky Mountain high. This whispered news, though, she calls a Death Valley low. She debated whether she should give her lecture as planned or instead deliver an impassioned plea to bombard senators with letters of SETI support.

"It wouldn't have done any good," she says.

Tarter usually accepts the boulders and the grades up which they must be shoved. But she for once accepted that another person's will could defeat her own.

❖

The next morning, before the debate began, she left on a jet plane back to California. The congressional conversation took place while she was in the air. Even cruising altitude was not quite high enough to give perspective. While she looked down at the clouds and flipped through *Skymall*, her father's voice came into her head. "I don't see why you couldn't do anything, if you work hard enough."

Maybe he had been wrong.

She ran to a phone as soon as the plane landed. "Are we okay?" she asked.

"No," a colleague said. "It's done."

Bryan had won. His press release, typed onto stationery mocked up to look like the SETI Institute's letterhead, was headlined SENATOR BRYAN ENDS THE GREAT MARTIAN CHASE. "As of today, millions have been spent and we have yet to bag a single little green fellow," the release continued. "Not a single Martian has said 'take me to your leader,' and not a single flying saucer has applied for FAA approval. It may be funny to some, except the punchline includes a $12.3 million price tag to the taxpayer."

❖

"Don't leave me alone with any sharp objects," Tarter said to Welch when she arrived home.

Just a year earlier, at the HRMS launch, she had been so hopeful, had thought such grand thoughts, had compared her team to *Columbus*, for God's sake. And now the dream was dead. She couldn't even push a boulder if she'd tried. All the boulders had, in fact, been summarily carted away.

To the world, though, she showed a stoic face. "This is an enormous setback," she said to the *New York Times*. "NASA has spent 20 years and more than $50 million to develop sophisticated digital receivers capable of listening to tens of millions of frequencies at a time. Now, with the observations getting under way, the project is killed."

Barney Oliver was less circumspect when he wrote for the science newsletter *Signals*:

> *Millions of transistors, memory cells, and other high-tech products of our ingenuity have been woven into a brain whose sole aim in life is to detect and verify the origin of tiny signals—less energetic than the smallest atomic particle— that have crossed the light years we cannot. Such signals will tell us that we are not alone, that the astonishing process that has produced us out of the fiery furnace of the Big Bang has also occurred elsewhere. Lo, from that single fact, all our philosophy would be enriched. To save the American Taxpayer about eight cents per year, we are to be denied the chance to explore the universe and the sentient life forms that fill it.*

It was the kind of oratory Tarter would later give. But that October she could only mope and avoid her knife block.

The next day, though, a call came from the targeted-search project scientist John Dreher, who had joined the team in 1989 after leaving a physics position at the Massachusetts Institute of Technology. "You know," he said, "if what we were doing yesterday made sense, it's still going to make sense on Monday. We just have to find some other way."

CHAPTER 8

THE LAST CHAPTER

The next Monday, Tom Pierson called everyone to the SETI Institute's boardroom, where he had piled fundraising books in the middle of the table.

"Everybody pick one," he said, pointing to the stack, "and read the last chapter."

The final section of any book about "institutional development," a financial euphemism, concerns the Big Ask—how to convince those with fortune and fame to bestow the former upon you so you and they can obtain the latter. Ideally, before reaching this section, fundraisers have gained a base of small-scale donors. Then, from those, they have culled and cultivated the richest, most dedicated patrons. And eventually they ask them for something huge—millions of dollars, the deed to their estate, their firstborn. "We didn't have time for all that," Tarter says. "We had to get money and get it right now." And so Pierson sent them diving into the denouement.

Pierson, who co-founded the institute and helmed it until a cancer-induced medical leave in 2012, was one of the few in the room who wasn't a scientist or an engineer. Development was his gig. Pierson's rounded nose and puppy-brown eyes did his character justice. He was kind and loyal to, even defensive of, his employees, qualities they reciprocated equally. And so Tarter and her colleagues hit the books, took their suits (some unused since their last-attended funeral) to dry cleaners, and practiced their pitches in front of mirrors. All except Barney Oliver: Oliver didn't need to practice. But when his colleagues began to make their shortlist of people who would maybe help government-rejected scientists find aliens, he held up his hand to stop them.

Oliver had more connections in Silicon Valley than any of them, because of his decades with Hewlett-Packard, and many favors he hadn't collected on, so Pierson put him in charge of contacting the potential donors. But Oliver said, "I can't talk to anyone until I talk to Bill and Dave [Bill as in Hewlett; Dave as in Packard]. Anyone I talk to is going to ask me what Bill and Dave are doing."

He secured them all an appointment.

❖

At the Hewlett-Packard offices, paintings of California countryside landscaped the walls: the lumpy golden hills, which resemble the humped backs of giant creatures; the snow-packed peaks of the Sierras, whose water is the only reason people can live here at all; fogged-over sea rocks straight out of a fantasy novel. Tarter watched them pass her by as she headed to the executive suite with Oliver, Pierson, and Frank Drake. They all shook hands with Hewlett and Packard and sat down.

Oliver gave the executives a speech about humanity's deep future and our blood-borne need to find out if we're alone. It's the ultimate question! But, he said, the government had recently decided it couldn't care about those most fundamental human impulses.

"Damn politicians," mumbled Packard—ironically, as he himself had acted as US deputy secretary of defense in the 1970s and whose company sold billions of dollars of equipment to the government.

Tarter then took over the presentation with the specs of the project itself—the frequency ranges they were searching for extraterrestrial transmissions, the sensitivity of the antennas, the stars they planned to search. She thought that even if she didn't give the slickest oratory, she at least gave an accurate one.

At the end of their speeches, Hewlett nodded a few times. "Thanks very much," he said. "We'll discuss it."

The meeting over, the group then ushered themselves back through the paintings of paradise. Oliver led his pack back to the office he still held at the HP Labs, putting his arm on Tarter's shoulder as they walked. She shrugged away from him; she hated it when he did that, and he did it often.

"Barney looked like my dad, and for me he was a father figure," she says. "But, boy, did I have difficulties with him. He could be just too patronizing."

Although she thought of him as a mentor, and was grateful to him for all the energy he devoted to SETI, the dynamics and gender politics between them too often tilted in his favor. The arm thing. The "which hunts." And he often dismissed her scientific arguments in a disagreement. While she sought the approval and support of a father figure, she also wanted to be seen as a mature expert and not a girl in need of permission and correction.

Tarter let go of some frustration with his attitude toward women when she thought about how she'd seen him treat his wife, Suki, at a NASA anniversary celebration three years earlier, when she began to show overt symptoms of Alzheimer's disease. After the illness had begun to vignette her consciousness away, he continued to bring her to such social functions; he patiently explained who all their old friends were, how to butter bread, what this or that banquet was for, and why she was welcome there. And later, he established the Priscilla Newton Theater Arts Scholarship at the University of California Santa Cruz to commemorate her interest in dance and the theater.

At the NASA celebration, Suki began to cry at dinner.

"Suki, what's the matter?" Welch asked.

Suki said, "I don't know who I am."

The statement rang through Tarter. Although she claims she spent most of life—until a few years ago—thinking of herself as immortal, the old-age loss-of-self did sometimes burst through her more-youthful consciousness. *Who are we when we don't know who we are?* she wondered. *If we are only our memories, what are we when they're gone? What is a self?* The questions spiraled like accretion disks. But she simply smiled and said, "You're Suki Oliver. You are married to Barney Oliver. You have three children. You were a really good community theater actress. We are pleased to have you join us for dinner tonight."

Summaries of self can't help but feel like eulogies. Where else (besides perhaps awards applications and dating profiles) do we list our most fundamental fundamentals, and what we will leave behind? And Tarter wondered, *What would someone say to me if I asked them who I was?* She was beginning to think about her legacy, without even realizing it.

As the SETI team approached the lab building, Oliver said, "I'll buy you all lunch!" he said.

But as soon as they arrived at Oliver's HP office, his phone rang.

"It's Dave!" Packard said. "We want to talk to you."

"I just promised I'd take these nice people to lunch," Oliver told him.

The ducklings quacked at him: *Go, go go! We can feed ourselves.*

Half an hour later, Oliver returned from HP triumphant: "We got our first $2 million!" he shouted. H and P were in for $1 million each, as long as the SETI team could find others to match that amount. Tarter's first solo fundraising success came in at $10,000 from Mitchell Kapor, founder of the software giant Lotus Development Corporation.

❖

Gordon Moore of Intel was the next bigwig on the list. Intel was a sprawling complex of offices, partitioned like farmland. It contrasted sharply—in its looks and how the business functioned—with HP, which was known for its humble start in a one-car garage. But Moore listened to Oliver and his entourage, sitting in his double-wide cubicle. After checking with his family for their approval, he personally wrote a $1 million check.

They next went up to bat against Paul Allen, the co-founder of Microsoft. When Oliver first called, Allen's administrative assistant said she didn't want to bother him with this pitch. He's busy, she said. He's not interested.

Oliver cleared his throat and spoke in what others called his gruff, corporate warlord's voice.

"You put this meeting on your boss's calendar," he said to her. "He can take it off if he wants to."

Allen didn't take it off, because Oliver was one of his childhood heroes. The meeting lasted nearly an hour and ended when Allen said he needed to talk with his sister, Jody Patton, who serves as the CEO of Vulcan, Inc., Allen's investment and project management company, and the president of the Paul G. Allen Family Foundation. Without any more prodding, Allen sent a handwritten check for $1 million. When it arrived, the team crowded around the rectangular paper, counting zeros.

The Big Asks continued. If Oliver couldn't get in a room with an exec to "have a conversation," he would offer more: "I'll rub your feet, massage your hands, whatever it takes." (No one ever took him up on that.)

The team eventually went back to Hewlett and Packard with their collection plate, demonstrating that they had matched the money.

Bill and Dave shelled out a million more each, bringing the total to $7.5 million, including the $100,000 Oliver himself contributed. It was enough to resurrect HRMS, no government necessary. "Given the fact that it wasn't possible to fight for it, I have incredible respect for Jill and her leadership, and I think she did a wonderful thing instead of giving up to move forward and turn it into a privately funded activity," says Goldin.

It was the last time financial support would come so easily.

❖

During the six months following the SETI project's "termination," as the scientists and NASA headquarters bureaucrats apocalyptically call it, Tarter went to the SETI Institute at 6:30 every morning. At a kid-sized desk in the corner of the boardroom, she sat as if in time-out, writing the "termination plan" that would allow the team to keep the equipment they had built. If they were going to use deep private pockets to continue the search for extraterrestrial intelligence, they needed to hang on to their instruments, which technically belonged to NASA. Tarter wrote pages and pages that cataloged every circuit and CPU, each component annotated with justifications of why it could only be useful to SETI astronomers and would be no good for anyone else at NASA.

Then, at 10 A.M. every day, Tarter drove down the road to NASA's Ames Research Center to work on that very equipment. And then at 4:00 P.M., she returned to time-out and the termination plan. Assistant administrator Wesley Huntress, who replaced Len Fisk in Space Sciences, helped smooth the transition when he took the position in 1993.

After many such days, the federal government agreed that the equipment was worthless to anyone but the alien hunters, labeled it "surplus," and donated it to the SETI Institute. JPL's brand-new wideband feeds and receivers went to Arecibo Observatory on an intergovernmental loan. The SETI scientists were also able to hang on to their beloved trailer, the Mobile Research Facility, on which technician John Ross had worked tirelessly. They kept their home. They could go anywhere, do—maybe—anything. Or at least keep doing something.

Except that without NASA's network of telescopes, they had to abandon the sky-survey component. And they had no connection to the military plane that could airlift the trailer to the telescopes. In a Hail Mary pass, Tarter wrote to Virgin Airlines' Richard Branson, the blond-bearded hard partier who today wants to ensure that the one percent can daytrip to space. Did he have any front-loading 747 jets? Virgin's slogan at the time was "Don't let anybody tell you you can't do something."

Tarter wrote to him, "Congress has just told us we can't do SETI." And now we need to borrow a big plane. The cannibalizing of his phrase (or something else) caught Branson's eye, and he wrote back, wishing them luck but informing them that Virgin had no suitable jets.

❖

When the SETI Institute resurrected their search for extraterrestrial intelligence, they called it Project Phoenix, a self-created mythology about rising from the ashes of congressional termination. Now all they needed was a telescope to which they could hook up their equipment and start searching. In the United States, the large radio telescopes—the only kind useful to SETI—are connected to the federal government, whether that be the National Science Foundation or NASA. The same is often true abroad. And because taxpayers support those telescopes, they are public, with some responsibility to "the people." Organizations can't just buy time on them—they need to be equal-opportunity, with proposals ranked by merit. For many taxpayer-funded telescopes in the United States, anyone (even you!) can submit their ideas and specifications and see how they measure up against those of other citizens. But, in the past, the committees ranking proposals and allocating time on the telescopes had put SETI toward the bottom of the list, or at least not high enough up to earn them observing hours. But soon, Tarter found a facility for which money did actually equal time.

While visiting Prague for a conference, she had a drink with the director of Australia's National Telescope Facility, Ron Ekers. After listening to her plight, Ekers leaned back in his chair and pointed his glass at Tarter. "I've got a telescope for you," he said. "We could rent you Parkes."

Parkes is the 64-meter radio telescope made famous by the movie *The Dish*, in which scientists use it to downlink data from the *Apollo* landing after NASA's engineers fail to do so, saving one of the most famous television broadcasts ever. While renting a telescope like this isn't unheard of, it's not often heard of. With costs today of up to $0.50/second, not many private groups have the resources, and not many telescopes offer themselves as commodities. But Ekers wanted to build a new kind of radio receiver—and the Australian government wouldn't help. But Tarter could. For $250,000, Ekers said, he would give her Parkes 24 hours a day for 6 months, and the Australian engineers could build the receivers she needed. The two shook on it, making an international scientific deal over pilsners. Tarter went home to await the contract.

But then the signed contract didn't arrive. Finally, the CSIRO lawyer called Tarter. "We can't sign this," he said. "Suppose Paul Allen decides to sue us if you come and use the telescope and then don't detect ET?"

His worry had a basis: an Australian farmer had recently sued the organization because of its weather-forecasting technology. He had used it to help plan crop placement, and it—and so he—had been wrong. And he'd lost his farm fortune.

Tarter (and Ekers, off-camera) explained to the Aussie lawyers that neither Allen nor any of the other donors had placed any "detection strings" on their gifts, and soon the contracts came.

❖

Because the MRF trailer had no air transport, it had to travel by sea, a slow journey that began months before the scientists flew over. They gutted it ahead of time, so they could fix and upgrade the electrical innards while the trailer floated across the sea. Those guts, left behind, later had to travel to Australia in the scientists' personal luggage. Although airport security wasn't as tyrannical in the '90s as it is today, 16-layer circuit boards, corrugated feed horns, and HP test equipment looked a lot like bomb parts. An International Traffic in Arms convention protects the temporary shipping of scientific equipment across international borders, even if that equipment looks like it might down a plane, and shields them from customs and duties. Tarter had written up the necessary inventory and believed she had all the right papers to bring the suspicious SETI electronics across international borders.

"We're just going to Australia," she thought. "We're not going to the People's Republic of China."

All should be fine; besides, they looked like trustworthy, Dockers-wearing engineers. But customs always detained them, and customs officials' gloved fingers clutched the corners of silicon rectangles, demanding an explanation.

❖

In January 1995, after they finally made it out of the bowels of the Sydney International Airport, Tarter and 20 engineers descended on the Parkes Radio Telescope. "I honestly don't know whether to call out the National Guard to protect you guys," said the station's head, Mark Price, "or just rent a whole bunch of Porta-Potties for the press and tourists."

Not a whole lot, besides this alien thing, was going on in the rural area around the telescope. The Parkes antenna rises from the farmland about 15 miles from the town of Parkes, surrounded by a baseball-field-like arc of golden grass. A diagonal road shoots from what would be the catcher's mound. And all around the field, more fields—these full of crops—grow in perfect, intelligently designed squares.

Tarter's entourage overwhelmed the observatory's infrastructure. Even bunked like college freshmen, the engineers couldn't all fit in the observatory's dorms. Some of them boarded in motel rooms in the town of Parkes, an 1870s gold-rush hamlet. But Australian drinking culture—and specifically a pub crawl orchestrated by the welcoming town officials—threatened to derail their tight schedule. One desperate young engineer showed at the observatory late the morning after, having prescribed himself a run from town (20 miles) as the best possible cure for his hangover. This was T-minus one month until Phoenix took flight.

Nothing was ready. After collecting the MRF from Sydney Harbor and hauling it overland to the telescope, they had to reassemble it. And, even more difficult, they had to get a second telescope—one called Mopra, 144 miles away in Coonabarabran—running as the host for a follow-up detection device (FUDD), meant to confirm that the signal came from space and not local electronics.

The scientists planned to have one FUDD at the main telescope, to analyze the signal intensely while the telescope continued on to a different part of the sky.

The scientists planned to observe simultaneously with the Parkes and Mopra telescopes, looking at the same star system at the same time. The Phoenix computer at Parkes identified candidate signals and compared them with a database of known human interference. The FUDDs at both sides then looked back at any viable candidates. And the ones that persisted at Parkes were compared against the data sitting in the Mopra FUDD's memory. A signal truly coming from the sky (and not from, say, a nearby airport) would look different at the two locations because of their different locations on the planet.

Tarter had planned to run Mopra remotely, from her perch in Parkes, but the "remotely operable" feature of the Mopra system wasn't working yet. Tarter and the team scrambled, working around the clock to bring that system, and the rest of their gear, online. It often fell to Tarter to phone the head engineer at Mopra in the middle of the night, and she cringed when his wife answered the phone in a sleep-thick voice.

Over the course of that month, Tarter did her best to turn Parkes into a place her people would want to live, or at least wouldn't leave. She rented a U-Haul, went to "this thing called Ikea," and bought DIY particle-board desks and assemble-it-yourself office chairs. She bought a candelabra to adorn consoles and a coatrack for their hard hats from a thrift store. Armed with a can of spray paint, she covered them in gold. She made filled glass containers with candy and made thrice-weekly runs to town for Coke and Tim Tams for the young engineers. It was, in a way, a home.

The coral-pink trailer Parkes gave them to use as an office had a sign: JILL'S ROADHOUSE. HOME SPECIALTY: DEEP-FRIED PHOENIX IN TARTER SAUCE. Some called it Jill's Palace. Those who did were mostly men, but most everyone was. The Australian men loved to make blonde jokes of dubious taste, with sexual innuendoes as punchlines. Tarter, sick of this, stared at them blankly after one joke's delivery. She blinked, willing vacuity into her eyes. "I don't get that one," she said. "Please explain it to me." They never told another blonde joke around her again.

While there, Tarter did meet some exceptional women, like Carol Oliver, who became the voice of SETI in Australia, and Bobbie Vaile, an astronomer who had worked with Frank Drake to develop an astrobiology curriculum for her university. For Vaile's birthday, they celebrated by eating a cake iced with a Pinocchio-nosed alien. But Vaile, they soon learned, had a brain tumor. Treatable, perhaps. Time would tell, as it always does eventually. Vaile was 15 years Tarter's junior, and the idea that an illness in someone so much younger than she could bring a life and a promising career to a halt rattled her. She tried to stay hopeful.

Tarter also befriended the Australian women who ran the Parkes dorm, some of them nearly as broad as they were tall. Every week, these women went to the town's RSL and Services Club, a gathering place for current and former military, and line-danced to songs like "All My Ex's Live in Texas." Tarter joined them one night and marveled at the precision with which they executed the intricate steps. None of them had Texan exes, but it was perhaps true that the things they loved and longed for lived far away.

❖

On February 1, the Phoenix was scheduled to lift itself out of those ashes, shake itself off, and fly off to grab aliens in its talons. The team planned to point both the Parkes and Mopra telescopes at *Pioneer 10* as the first test, as usual. The spacecraft was billions of miles away, a tiny metal box of circuits in space, pinging away at nothing. They hadn't seen its gray diagonal signal since Arecibo. Before that, during HRMS, they'd seen it every day.

The morning of the test, prior to the arrival of the press and dignitaries, was less ceremonious than the Columbus Day beginning of the HRMS project. The neighboring farmer had started to fertilize his fields, and the winds had carried the smell to the observatory. A staff member was dispatched to get the farmer off his tractor and invite him to lunch.

The team was slightly out of the spotlight. Even if they thought they were doing grand things, they now understood that there might be value in keeping that to themselves. They hovered over the five CRT monitors and CPUs that displayed their data and controlled the instruments. Tarter told the telescope to slew to *Pioneer*'s position, and they all held their breaths as they waited for a signal to rise up from the static.

Tarter thought of all the Tim Tams they may have eaten in vain, and how much they missed their families and friends, and how maybe this would just end like every other SETI project and . . .

. . . and then *Pioneer* slashed across the screen. The FUDD at Mopra confirmed. They had found evidence of themselves, of all of us.

❖

Most of the engineers returned home, but Tarter and the scientists stayed at Parkes for six months, babysitting the algorithms and correcting anything that was needed. Tarter often worked the late shift, a period that has always suited her, at least based on the time stamps of her emails. Every midnight, she walked 1.5 kilometers from the dorm to the telescope and sat under it till dawn. "I got so attuned to country living and the pace of life," she says. "I started my shift under this beautiful sky and hardly ever saw any people."

When, after months of that slow-strolling existence, she had to travel to the town of Dubbo with its 40,000 inhabitants to buy a transformer, she felt like a recluse. "I thought, 'Who are all these people? Get away! You're too close.'"

After Phoenix's successful Australian deployment, Hewlett, Packard, and Moore all pledged $1 million more each for five years. A science trust fund.

❖

Just after her sojourn Down Under, Tarter traveled to Star Island, New Hampshire, in July 1995 to attend a conference on the links between science and religion. Tarter didn't really see any. "Somebody twisted my arm or leg or something," she says.

Star Island is part of the Isles of Shoals, a chain of landmasses six miles from the East Coast, bleeding across the border between Maine and New Hampshire. Star Island is one of the largest at 46 acres. Small rocks ring the island, looking from the air like a dusting of snow, and all the buildings are perfectly manicured New England white Colonials. One night of the conference, Tarter sat on the dock with a Buddhist monk who told her that he always dressed in his robes for flights because the airline was more likely to upgrade him. Free drinks, more legroom, and fashion seemed like worldly concerns for a man of the cloth. But she liked his honesty. They watched the sunset together, the kind of sunset that makes you feel lucky to be an Earthling. That moment may have been forgotten if the next few days hadn't cast it into what artists would call "sharp relief," etching it deeply in contrast to what came after.

When she left the island to meet Welch and his daughter in New Hampshire, she spent a night alone in a Massachusetts hotel room. And out of nowhere, she felt a sharp pain in her armpit, like a letter opener shoved deep into the muscle. Her skin seemed to clutch inward toward the feeling—or, perhaps, toward a *thing*. A growth.

She called her doctor back in California. The phone rang and rang. He had left town for a few days, the receptionist said, and would call back soon. Tarter shoved the pain aft in her mind and packed for the next leg of the journey. She tried not to think of the word *cancer*. But trying not to think of something is humans' worst skill. We can build skyscrapers, write sonnets, bend silicon to our will, and search for aliens—but we can't *not* think of something.

When the doctor finally called back, he said, "Don't worry. Cancer doesn't usually show up as a sharp pain."

❖

The trip continued, with just the stab and tug beneath her skin, and the self-soothing of the sentence "It's nothing; it's nothing." When they arrived home two weeks later, she went to the doctor's office. The waiting room was a purgatory that felt like hell. So did the days that came just after that, before the diagnosis: *breast cancer*.

The treatment began right away—cut, poison, burn. Surgery, chemotherapy, radiation. The cancer was aggressive, so the treatment rose to match. Appointments listed in her calendar, once back-to-back committee meetings, become Tetris-style medical appointments: surgery, bone scan, CT scan. Entries like "SETI session" are crossed out, replaced with "blood tests at 4" and "chemo at 2." Still, she worked. "Even during that period, she was kicking everybody's butt," says Harp. "Even in chemo, she was still working harder than anybody."

Her hair fell out in clots. "I looked at myself in the mirror, and I said, 'I know that face,'" she says. "I looked like the bald baby-picture version of myself."

One of the other calendar entries from that period was "Barney's Memorial." While preparing Thanksgiving dinner, Oliver suffered a massive coronary, dying immediately, just weeks after Tarter had delivered a report on the Australian Project Phoenix campaign to him, Hewlett, Packard, and Moore. At a tribute service years later, Tarter paid homage to her substitute father figure and mentor before an audience of many hundreds crammed into the First Presbyterian Church in Palo Alto, California. She tried not to sugarcoat his larger-than-life personality, while saluting his massive intellect and the kindness he had shown to her by allowing her to stand on his gigantic shoulders. "Since the fourth of August, our cosmos has been 'out of order,'" she said. "That was the day the cosmos and all of us lost Sir SETI."

When she describes this time of her life to me at her house in 2015, twenty years after it happened, she veers away from emotions as quickly as she steers toward them, lapsing instead into serious science mode. For instance, when discussing the cancer, she says, "Histologically, the cancer cells were quite distorted. Disturbed and very . . ." She trails off and lifts the photo album page by its corner.

The next set of plastic-sleeved images shows her mother's house. Betty, a woman with a curled gray coif, the kind you "have done," sits in an armchair. Tarter has a straight-haired blond wig plopped on her head. She looks a little like one of the Beatles. "I hadn't told my mother about the disease at all," she says. "My father had died of cancer, as well as my mother's sister, my favorite Aunt Helen. It was the big C-word in our family." Instead, she told her mother that a salon treatment had gone awry. And when her mother asked her to install a ceiling fan, she did it, despite how she could hardly move her left arm after the surgery.

The treatment continued for nine months, from July 1995 to March 1996. She never told her mother. "She may have known," Tarter says. "She *never* let on to me. We never had a discussion about it."

It's human nature not to want to worry our parents. Despite how we behaved as teenagers, we know that our troubles and behaviors don't just have merely Newtonian impact on them. The reaction can be much greater than equal. They knew us when we really *were* bald babies. In some ways, we will always be bald babies to them.

Some families stage political debates at Thanksgiving, know details of each other's love lives, and actually talk about their feelings. Other families bury their troubles into the house's foundation. They use neutral tones or reserve the right to remain silent. They switch the topic to genetics when the limbic system gets too active.

❖

Toward the end of Tarter's cancer treatment, she went back to Parkes. When she arrived, she found that Bobbie Vaile was dying. Vaile's cancer had grown beneath her skull like it wanted to replace her brain. And that's the irony of cancer—it bites, clean off, the hand that feeds. Vail knew the jaws soon would clamp down. So did Tarter.

To see someone else dying, and to know the same disease wants to take you, too—it must be like practicing saying goodbye to yourself.

"My hair grew back in curly," Tarter continues, steering away again. "I assume that the DNA gets altered because of the chemicals, but then it reestablishes itself."

After returning from Parkes, she threw away the wig and walked around like a newly shorn sheep, little shoots sprouting back up. She got to have another chance. She stuffed down her goodbye, the grief over things she didn't want to leave, and her idea of what she would leave the world. She went back to acting as if she were immortal.

A few years later, the C-word came for Tarter's mother, just as it had for her father. Betty's cancer sprouted and spread like kudzu.

Tarter flew to Florida, dismantled her mother's life, and sold it to the businesses there that specialize in end-of-life auctions. She took her mother back to Berkeley with her and spent the next three weeks with hospice at the house. When she describes the period, she says only, "Hospice is hard on everybody. It was a sad time. I wasn't very skilled, and I still chastise myself over my incompetence."

A few months after Tarter's mother died, she flew with her mother's ashes back to Belle Mead, New Jersey. "I buried them next to my dad," she says. "I didn't know what else to do. I don't believe in an afterlife. My mother had been a widow far longer than she had been a wife, and I thought it would be appropriate to have a commemoration in stone of their time together."

❖

Phoenix had many lives, and began a new one in 1996 when Tarter's team packed up the gear in the MRF (trailer) and set off for Green Bank. SETI traveled back to its ultimate roots, all the way from Australia to West Virginia, where Drake had done his first experiment in the 1960s. Project Phoenix used the 140-foot telescope, which looks naval on the outside and steampunk on the inside. It has a giant white base that resembles a beached ship, and the arc on which the telescope turns calls to mind a captain's wheel. But if you climb up the ship's ladders and open its starboard doors, you'll find gears and gears and gears, lubricated with great drops of hydraulic oil that make the floor slick. The telescope turns on the world's largest ball bearing, more than 17 feet across, which shipped to the observatory by railcar, but narrowly—only four inches to spare between its edge and the railroad tunnel into Deer Creek Valley. Pulled from the train station to the observatory on a lowboy, it bottomed out crossing a curved wooden bridge, and had to be hauled off by a tractor, once the bridge had been greased with that slippery hydraulic oil.

The confirmation telescope, which would become home to the second FUDD, lived hundreds of miles away in Woodbury, Georgia.

AT&T had donated twin 26-meter dishes to Georgia Tech when it no longer needed them. Tech's astronomers Paul Steffes and Dave DeBoer agreed to work with Tarter to get one in shape for SETI follow-up. The first order of business: connect it to the outside world. AT&T hadn't wanted competitors take advantage of their abandoned technology, so they had cut all the cables at ground level.

So together, Project Phoenix and Georgia Tech rebuilt one of the dishes. The first time they tried to move it, the sound of metal on metal slashed across the landscape. The engineers came out of the control room and looked up at the gear that moved the telescope. One of its huge teeth lay on the ground.

"Goddammit," DeBoer said.

To continue the project and fix the instrument, they stole a gear from the other telescope. The FUDD, now a Frankenstein's monster, did well its follow-up job—except for a few days in 1997, when they really needed it. Then, the Woodbury telescope lay still and silent, with a dead disk drive after a lightning strike.

❖

In the early hours of June 24, 1997, Tarter paced back and forth along the narrow tile of the Green Bank 140-foot telescope's control room. On one side of her, the control panels, which look like they belong in a clunky 1970s starship, hummed. On the other side, the Phoenix computers churned a strange signal around their circuits. The telescope had zoomed in on the star YZ Ceti, a dwarf that regularly flared with bright bursts of radiation. But the equipment picked up more than those bursts: it also found a steady signal, arriving at a particular set of frequencies that were equally spaced across the dial. It was just the kind of radio-station-style broadcast from space—from intelligent extraterrestrials—that Tarter dreamed of, and exactly what they had designed their equipment to detect. Such a signal exists (as far as we know) only when someone creates it. *Someone* could mean either humans or extraterrestrials, though. So before Tarter could break out the Champagne, she needed to make sure the team hadn't made the less monumental discovery of an airport radar or a North Korean spy satellite. Had the Woodbury telescope been alive, they would have done the usual follow-up, seeing if the signal showed up the same there. Instead, they had to improvise. Tarter roused John Dreher from sleep in the dorm, and together they commanded the telescope to move far away from YZ Ceti, to see if the signal disappeared. It did—that was a good thing, indicating it was really coming from a single spot in the sky. When the telescope was pointed back on the star, the signal reappeared. They repeated this cycle multiple times with the same results. Then, at Dreher's suggestion, she tilted the telescope just a tiny distance away from the star. The signal stayed just as strong.

It didn't make sense, said Dreher. If the signal were actually coming from the star, as the first sequence of tests indicated, it should have dimmed by a small amount in this new test.

By then, it was late afternoon and the star would soon set. Tarter debated calling another observatory, located to the west, where YZ Ceti would still be above the horizon. But other astronomers didn't like having their observations interrupted—what if they missed some truly exciting blazar blast or neutral-hydrogen action because they had decided to check up on a dubious claim that ET had finally phoned home?

Besides, Tarter didn't want to shout, or even whisper, about that potential phone call in public yet, not until they were sure. A false announcement could ruin them; crying wolf never helped a struggling cause. That is, after all, the point of the fable.

She had, though, called California, and her colleagues at the SETI Institute were watching a mirror-copy of the Green Bank control room computers that sat in their lobby. Back in West Virginia, when YZ Ceti set, Tarter and Dreher turned the Phoenix controls back to automatically search the next stellar targets rising in the East. They gathered up the log files and printouts from the day's activities and headed off to dinner, with a growing sense of disappointment that this signal was not the One.

Sometime during the day, the *New York Times* called Seth Shostak, Tarter's colleague and a favorite of the media because of his down-to-Earth dad humor. Science journalist William Broad wanted to know about this signal they were tracking down there in West Virginia. How had he heard about it, in those days before social media, when the team had locked themselves in the control room? Broad explained that he was working on a tribute piece for Carl Sagan, that would post on the second anniversary of his passing. He had just called Sagan's widow, Ann Druyan, who had that morning talked to Tarter's unflagging administrative assistant, Chris Neller, who for decades kept the institute—and Tarter—running from behind the stage curtain. Neller had told Druyan that Tarter had delayed her homebound flight to "take care of some business" at Green Bank. In SETI, that's suspicious.

Shostak, according to his book *Confessions of an Alien Hunter*, told Broad, "Well, we're continuing to track the star. But, you know, these things often turn out to be man-made interference. We're checking out a lead on that right now. But I think we'll know more in three hours or so. Can I call you back then?"

Broad said yes, probably sure he had the scoop on everyone else because of his fortuitous phone call.

By the time YZ Ceti crawled above the horizon at 5 A.M. the next day, Tarter and Dreher's perusal of the previous day's data and a bit of Web searching had turned up a mundane match for their signal: the Solar and Heliospheric Observatory (SOHO), a satellite that takes close-ups of the sun that make it look alive and angry. Every star looks like that, with a campfire-coal surface and loops of plasma protruding from its surface. Maybe somewhere else in the Milky Way, another civilization knew how to take such pore-level portraits of their star, too. But the point was, they had found—once again—only evidence of us. Unfortunately, they had forgotten to tell their colleagues back in California about this mundane explanation, and those folks had stayed in their offices late into the night, nervously waiting for YZ Ceti to rise and the search for the signal to begin anew. A significant amount of professional fence-mending was required when Tarter returned.

If the Woodbury telescope had been functioning, if the spacecraft had been orbiting Earth instead of the sun, and if Tarter had not misread the output of a database query she made shortly after the initial discovery, this incident would have resolved itself right away. Instead, they learned a lesson: If they ever *did* detect a communication from an extraterrestrial civilization, they wouldn't be able to keep it a secret. The William Broads of the world would find out fast—and this before Twitter.

❖

No more alarms—false or otherwise—rang while Phoenix operated in Green Bank. It traveled next to Arecibo, the jungle where its predecessor, HRMS, had first begun on Columbus Day six years earlier. Harp says that during their long evenings in the jungle, Tarter would turn on samba music and dance around the small area off the main control room, where the Phoenix computers issued controls to the telescope and displayed any signals being tracked. All the Phoenix detection equipment lived in the parking lot just outside, within the MRF trailer carefully pulled up the winding mountain roads in Puerto Rico. But this was a transformed MRF, with the old NASA-derived electronics ripped out and new PC-based processors slipped inside. It sat next to the control room, which is a ship captain's wall of glass that looks out over the bowl of rock and greenery in which the Arecibo Radio Telescope rests. Astronomers like to jog around the dish, which has a perimeter of more than a half a mile. Under its speckled shadow (holey from the mesh surface), jungle plants grow. Iguanas saunter up to the patio where the astronomers eat lunch. But above the telescope, something new hung like a chandelier.

Called a Gregorian dome, it looks like the Epcot ball with the bottom third cut off. It hangs 500 feet above the dish on an arced track, and scientists can direct it to travel along this track to aim at different parts of the sky. Before that, Arecibo was more limited in what it could point to and detect, and the upgrade made the telescope a better fit for SETI work.

The Gregorian apparatus attaches to a flat circular track, parallel to the ground. Along that track, the arc and Gregorian can twirl like Disney World teacups, allowing even more pointing. Most importantly for Tarter and Project Phoenix, though, the Gregorian contains two more shaped reflectors inside to correct distortion from the primary dish, making it useful for a wide range of frequencies. And that, in fact, is why NASA—preparing for its SETI program, back before Congress banned SETI from NASA—had invested the initial millions into the upgrade years earlier.

That the SETI got time on the telescope at all was a political favor. The cost of the NASA-SETI-funded contribution to the upgrade almost exactly matched the operational cost of the hours of observation the team requested. And although Arecibo is like other large-scale American facilities operated by the National Science Foundation, in that allocation committees rank proposals by merit—a process that doesn't usually go well for SETI—Tarter suspects the observatory director chose his proposal reviewers "carefully." SETI became the largest single user for the next 5+ years, taking up 5 percent of the telescope's total operation time.

A controversy around SETI taking over telescope time comes up in the movie *Contact*, which Robert Zemeckis began filming around this time. In *Contact*, Tarter's arch-nemesis David Drumlin tries to convince Ellie Arroway to stop doing SETI, because of its null results, its needle-in-a haystack-ness, and its lack of contribution to larger science—or even its subtraction from larger science, whose time it steals. Drumlin's speech doesn't work.

When fictional Arroway stays at Arecibo, she lives in one of several cabins that scientists do actually occupy when visiting the observatory. Dubbed the visiting scientists' quarters (or VSQ, because astronomers love acronyms), they sit high on a hill, up seven flights of outdoor wooden stairs away from the control room. Strange slippery molds and algae grow on the stairs, easily landing you in a pile of mud. The cabins have tiny linoleum kitchens connected to living rooms of bamboo furniture with floral, almost three-dimensional cushions. Two rooms with two beds—one of which is meant for you to share with Matthew McConaughey, if *Contact* is an indication—are nearly open to the elements. The sounds of *coqui* frogs (which sound like their name) tunnel through the wood walls. From the porch, you can look out over the kinds of lush trees whose oxygen production you can almost feel. The air tastes new.

Tarter has lived in these VSQ for many months total, including during Phoenix. Jodie Foster merely played at doing so during the filming. But when Foster took on the role of Ellie Arroway, she wanted to find out what being an astronomer really was like. For that, she turned to the person on whom Sagan had modeled her character.

Before filming began in 1996, Foster called Tarter regularly to talk. "It was never about the science," Tarter says. "She was clear that she wasn't going to teach anyone science, but she wanted to find out what the life of a scientist was like."

One day, Foster asked, "Do astronomers have egos?"

Tarter, who openly does, laughed. The answer is simple: yes. They're the surgeons of the physical sciences, godlike because of their cosmic dealings, or like prophets sent to interpret cosmic truths for the masses. Tarter laughed some more, to make sure Foster knew her response was sarcastic. "No," the radio astronomer said. "Well, maybe the infrared astronomers."

❖

One day during filming, when both women were doing their respective jobs at the world's largest telescope, Tarter took Foster up into the Gregorian dome. They rode the jerky cable car up to the support platform and trekked across the catwalk. Tarter never thought about falling, never felt that jolt when she looked down from a great height (instead, she says that the Gregorian platform offers the best views on the island). When they reached the middle, they ducked under the Gregorian's opening. Inside, the air felt cooler. Tarter looked up at the polygon-panel ceiling and listened to the *chirp, chirp, chirp* of the cryogenic vacuum pump echoing off the white walls. Right then, the telescope was collecting radio waves from many light-years away. These waves had left their home in outer space years ago. No one could see them, but the telescope's electronics turned them into pictures and plots that humans can make sense of. This dome let scientists see an invisible universe.

"This is an astronomer's cathedral," she said, turning to Foster. Then, switching back to science mode, she said, "It works the same way as the whispering dome at St. Paul's."

There, sound waves from human voices slink near the curved wall's circumference, arriving on the far side of the building nearly as loud as they were when they left. If an intelligent being stood across such a room from you, trying to communicate, you would have no trouble hearing their message, no matter how quietly they spoke. You would just need to be in the right place at the right time, listening.

CHAPTER 9

EXTREMOPHILES AND EXOPLANETS

T oday, the SETI Institute's offices are situated strategically in Mountain View, California, in the ever-expanding heart of Silicon Valley. From here, Tarter and her team have access to the tech world's best minds and their big money. Until 2015, the institute shared their two-story duplex building with a company called Jasper, which bills itself as "The ON Switch for the Internet of Things." The building itself is simple, gray. And at the back of the oak-shaded parking lot, a neighborhood-style basketball hoop hangs over prime shady parking spaces. Some guy does tai chi out there every day at lunch.

Tarter's office sits in the middle of the building, filled with paraphernalia: a director's chair that says JILL TARTER PRODUCTIONS, 22 different certificates, a model dinosaur skull, three globes, an autographed photo of Jodie Foster, and that screensaver: "So . . . are we alone?" It's almost as a movie set designer to constructed an office that Jill Tarter might have in a Jill Tarter biopic.

Around the office, which Tarter still visits weekly even though she's nominally retired, scientists study all the topics that relate to some factor in the Drake equation—that piece of philosophical math developed more than 50 years ago—and breaks down the ingredients that go into the rise of smart, communicative, off-Earth life. In the 33 years since the institute formed, the scientists here and around this whole planet have been getting more tangible results about the possibility that extraterrestrial life might exist—and not just in the usual binary-form answer to the question "Did we intercept a message or not?" Those tangible results come from two main fields: exoplanets and extremophiles.

Since the institute began in 1984, astronomers have discovered two important things: thousands of worlds outside the solar system and increasingly extreme life right here on Earth. The yield of that chemical reaction is the discipline of astrobiology—what scientists call the study of planet habitability, the conditions that cause life to arise, and the search for signs of life (intelligent or not). Astrobiology thus joins the two "ex-" fields together. And by embracing the spectrum of possibilities, from prebiotic chemistry to electromagnetic broadcasts, as all part of the search for life in the universe—something not all scientists do—Tarter has helped grow the institute to nearly one hundred chemists, biologists, computer scientists, astronomers, philosophers, and geologists—a far cry from the single-person endeavor SETI was when Frank Drake did the first Green Bank search.

Some of them work on star formation, others on planet formation and location, and still others on what makes a planet livable. Some dig in to what makes biology come about, some how biology becomes smart and what "smart" even means, and some how we could communicate to that hypothetical smart life and what *its* communications to us might mean or look like. The part of the SETI Institute that is actually dedicated to traditional SETI—searching for the electromagnetic missives from extraterrestrials—is much smaller than the other branch: the Carl Sagan Center for the Study of Life in the Universe, which covers all the Drake equation factors up to "What fraction of life becomes intelligent?"

All of that more traditional science informs SETI's likelihood of success. And lately, the odds are looking up, precisely because of how many—and how varied—scientists have found non-Earthly planets and Earthly life to be.

The most interesting of that earthly life comes in the form of so-called extremophiles. Extremophiles are exactly what their name indicates: life that loves extremes. These are the X Games competitors of the biological world. It almost seems like they are trying to one-up each other with weirdness. Some thrive on battery acid, others in the cooling pools of nuclear reactors. They love hugely hot places and really cold ones. The most famous and beloved species, called a water bear or tardigrade, can even survive the vacuum and radiation blasts of space.

Humans once thought life was a rather fragile thing, touchy. It needed just the right Goldilocks conditions to be okay. But scientists have discovered that what's "right" for the life that's most like us is actually wrong for life very different from us. As we grow up, we learn that in a cultural and sociological sense. But science was slow to catch on to that same truth in a biological sense.

Earlier last century, even the best biologists thought life needed water, oxygen, and sunlight, just as you learn in elementary school when you're trying to grow a lima bean in a Dixie cup. Flora and fauna could survive cold, but not too much; heat, but not too much; salty water, but not too salty. It couldn't hang in highly acidic or very basic places. Its DNA needed the shield of a nucleus.

We were so wrong!

Cracks in that thinking appeared first in the early 1900s. People noticed that their stores of salted cod were going bad. What they didn't understand was that there were scores of living microscopic animals that survived the salting process and were spoiling their food. Then, in the 1940s, miners found microbes in the seemingly toxic drainage at the Iron Mountain mine. Twenty years later, microbiologist Thomas Brock discovered bacteria in Yellowstone's otherworldly hydrothermal features—thriving in the hot, acidic environments that bubble up from the innards of Earth. In 1974, NASA Ames scientist R. D. MacElroy gave all of these strange beasts a name: extremophiles.

But "black smokers" really clinched the concept that life goes on basically everywhere. In 1977, scientists and explorers Jack Corliss and Robert Ballard found bacteria around thermal vents on the ocean floor, which pour steam up as if from Dante's depths. But the bacteria didn't care that these vents were both brutally cold and unimaginably hot (along with being pitch dark). Nor did they care that no one had noticed their hardiness. They just went about their bacterial business, filling their evolutionary niche like they always had.

Since then, scientists have unearthed even more extreme organisms: endoliths that eat the rock under Antarctic ice, methanogens make methane gas under ice in Greenland, Arctic cyroconites that have natural antifreeze proteins, *Thiomargarita* bacteria that slurp sulfides off Namibia's coast, bacteria like nitrosomonas that take in chemical energy instead of sunlight, and tardigrades even survive in space sans air or water or the pressure they provide.

With this rainbow of bad living situations that other beings don't actually find that bad, scientists have been forced into a kind of habitability relativism, in which we can't project our preferences onto other organisms—a good practice in general. Our optimal happy place is other life's (usually very small life's) oppressive environment, and vice versa. The diversity of livable real estate on Earth demonstrates just how many habitable spaces could exist in the universe, if we don't limit our search only to the neighborhoods exactly like the ones *we* would personally want to inhabit. So now, scientists can look for myriad conditions (and keep their eyes out for even weirder ones) on other planets or big moons, knowing that the possibility of finding a tiny creature peering back is not totally out of the question.

As Tarter said in a July 2014 talk at NASA Ames Research Center, "*Homo sapiens* is just one single leaf on a very expansive tree of life, and that tree is really densely packed with organisms that have been finely tuned over millions of years to meet their specific survival needs. Although we know that, I think a lot of our fellow inhabitants of the planet certainly haven't internalized this idea. Our egos haven't yet caught up to this scientific understanding. This is a perspective that the natural universe does not share. So get over it."

❖

When we think of potential peering creatures within our own solar system, we usually think first of Mars. People have hypothesized habitation there since the late 1800s at least. And thanks to probes and rovers like *Curiosity*, *Spirit*, *Opportunity*, *Viking*, and other strongly named robots, scientists know of water ice that sits on the planet's surface, liquid saltwater that flows down dunes, fertile soil, and ancient oceans. Mars may have a boring landscape and no good trees (or bad ones, for that matter), but it has—or has at least *had*, in the past—conditions that terrestrial beings could survive. That's why scientists are still looking for life there, despite having been disappointed twice before.

The first blow came in the 1970s, when two *Viking* landers ventured to Mars, taking along with them four life-detection experiments. Safely on the surface, four experiments tested for the presence of organic compounds and metabolism—ones that contain carbon (which, as far as we know, *is* actually a thing necessary for life). Three came up empty. But one, called the Labeled Release Experiment, did provide a positive result: it seemed to have detected some organism taking in energy and expelling waste (just like you!). Together, these experimental contradictions—in which instruments found no carbon, but one perhaps found beings eating and excreting—were deemed "inconclusive." The inconclusive conclusion that still holds today, although a study from 2010 suggested the experiments may have had no chance to detect carbon, even if it were there: the salts that spike Martian soil may have destroyed any extant organic compounds when scientists heated them up.

Still, nothing in the experiments, especially not a something, screamed, *"Life, for sure!"*

The second dashed hope happened in 1996, when a team of American scientists scouring for meteorites found one from Mars. It had formed around 4.1 billion years ago, when both Mars and Earth both were still "young" planets. Mars was nicer back then, hadn't been hardened and stripped of its atmosphere by the solar system's hard knocks. It still had liquid water sloshing around its surface. But around 17 million years ago, the piece of rock—fondly known as ALH84001—shot from Mars into space. After whizzing around and past who knows what, it eventually fell to Earth 13,000 years ago. It only took those scientists 12,980 more years to find it.

And when astronomer David S. McKay of NASA stared closely at the rock with his scanning electron microscope, he saw some-thing strange: fractalesque patterns. He'd seen the same patterns on this planet: they were fossils, from microbes pressing themselves against the rock. These meteoritic squiggles, he said, were evidence of ancient Martian biology—the Red Planet had been alive. McKay published a paper saying as much in *Science*. The world lost its mind, and as evidence of that, I present you with this: President Bill Clinton went on television to announce the news to the American people—and if you know how often presidents call press conferences about science, you know how special this moment must have been.

But ALH84001's fame was not to last. Other scientists showed that they could create such "fossils" chemically and geologically in the lab. And while that doesn't mean the meteorite *doesn't* have fos-sils, it did mean the meteorite doesn't *necessarily* have fossils. It could be life, or it could just be a geochemical process, albeit a Martian one. While McKay hung on to his conclusions, most—finding no other evidence in support of his hypothesis—moved on. But there's a lasting legacy of sorts from the scientific frenzy: So many scientists wanted to study so little rock that it became necessary to invent or perfect tools for studying small samples. Those tools live on and continue to be improved upon and miniaturized for future studies on other worlds.

Other places in the solar system make even better microbial real estate than Mars does. Saturn's largest moon, Titan, is about 160 degrees Fahrenheit colder than Earth ever gets. But it has a nitrogen and methane atmosphere, and lakes and oceans made of methane. While that might sound awful to you or me, there are organisms here on Earth that thrive in methane environments, so it would follow that there could be extraterrestrial methanophiles, too.

Europa, one of Jupiter's satellites, seems to have liquid water oceans beneath its icy exterior, pressing up against it just like the water of Antarctica presses up against its ice floes. Perhaps, at the bottom of all that, Europa has its own version of black smoker beasts, or sheets of algae-like biomasses clinging to the ice's underside, or animals that feed on the energy from the planet's gravitational flexing (if you have seen the movie *Europa Report*, you might also have some more sinister Galilean aliens in mind). Many scientists, including some at the SETI Institute, hope to send a probe there to find out.

❖

Beyond our solar system, sending probes isn't really practical at the moment. But in the past 25 years, scientists have discovered that many, many, *many* planets do exist out there—more planets than stars. This is a huge philosophical shift for science, scientists, and the rest of us.

It wasn't long ago that scientists thought other planets might not exist at all beyond the worlds within our own solar system, let alone be habitable. Around the time that the *Cyclops Report*, the SETI bible, came about and Tarter joined SETI, astronomers didn't yet know if hypothetical aliens would have anywhere to live. "I can't imagine what it's like to finish graduate school and begin this search, up against a complete abyss of knowledge," says exoplanet astronomer Debra Fischer, who now heads a planet-hunting project called EXPRES. "There's nothing to hang on to. What kind of courage it takes to look!"

And first, before the field focused on aliens themselves, it made sense to look for their worlds. "The first thing that we asked ourselves was, 'Oh, well, are there any other planets out there?'" says Tarter.

At that time, scientists hadn't abandoned the idea that perhaps planets formed when two stars passed close to each other and one stole a string of material from the other, pulling it out like a thread from a sweater. "If that were, in fact, the correct explanation, planets were going to be really rare," says Tarter. The competing model—the one that holds sway today—says planets form from the disk of material left after a dense portion of a giant molecular cloud collapses under its own gravity to form a rotating protostar. In this disk, small bits of dust and rock smash into each other and snowball, eventually, into Jupiters and Mercuries and—perhaps sometimes—Earths. This cosmic billiard game lasts until the protostar turns on and its radiation and particle winds sweep away any of the remaining disk. The planets left behind continue to interact, as the final architecture of the planetary system is established.

But scientists, being scientists, knew that the only way to tell how rare or abundant planets are was to set about finding some. But how?

Astronomers' first attempts used the precise position of stars in the sky. By mapping their tiny movements from our perspective, a method called astrometry, astronomers thought they would perhaps be able see the tiny tugs of planets pulling their stars around using gravity. But the precision needed for these measurements wasn't really achievable at that time.

Another method, called photometry, watches the light from stars to see if they dim on a regular basis. Just as you can pass your finger over a bulb and block out some of its light, so too do planets when they pass in front of their stars. Astronomer Bill Borucki began working on ways to make this method reality. His quest that would eventually—after much fitting and starting—lead to the Kepler Space Telescope and an embarrassment of exoplanetary riches. But not until decades after he first dug into the idea. A space telescope was needed to avoid the atmosphere's blurring effects, but doing things in space takes a long time and a lot of money.

Instead, a third method dominated until the *Kepler* spacecraft actually launched in 2009. It's called a radial velocity search. In this kind of hunting, which Fischer's EXPRES uses, telescopes watch for the see-sawing of a star as its planets tug it around, just as they do in astrometric searches. But instead of looking for a change in the star's actual position, they look for a change in the light waves. When the star moves slightly toward Earth, its light waves get squished, becoming bluer; as it scoots back away, its light waves get spread out, becoming redder. You can hear this phenomenon in audio form when you listen to an ambulance go by: its siren pierces at a high pitch as it moves toward you, then *wah-wahs* down in pitch as it passes by.

As happens amazingly often in astronomy, the first planet anyone ever found for sure didn't come from any of these methods. And it lived around a star, along with two other planets, that surely wouldn't support life: a pulsar. A pulsar is actually a dead star, an ultradense hulk of neutrons left after a supernova explosion. Astronomers Dale Frail and Alexander Wolszczan were watching pulses of radio waves coming from the pulsar B1257+12. These pulses should have pinged the telescope at highly regular intervals like the *tick-tick* of an atomic clock. But the scientists noticed that something was pulling the pulsar around, delaying the pulses sometimes and sending them to Earth early at other times. It was, they realized after much deliberation and data diving, a planet or three.

Holy shit, how did these things survive the explosion of the star, and/ or did they reform from the debris? Tarter recalls thinking.

But the first planet that reminded us—a tiny bit—of our own came only two years later. At the time, Tarter and the SETI team had just returned from searching for signals from intelligent extraterrestrials with the private Phoenix Project in Australia for six months. On an October day in 1995, Seth Shostak came in to their newly reoccupied office at the SETI Institute and said, breathless, "They have a planet."

It was a massive world in a 4-day orbit around its sun-like star.

"It was such a startling result," says Tarter. The world needed to expand its vision of what a planetary system could be—making this the best kind of scientific discovery, one that reveals something utterly unexpected.

The very next day, Tarter received an email from astronomer Phil Morrison, who had written that very first paper, from decades earlier, about how extraterrestrials might communicate. He and his wife were in the Southern Hemisphere at the time, teaching in Africa. But even there—and before the real Internet—word of this new world had gotten to them.

Since those heady early days, scientists have found thousands more planets, many of them with the Kepler Space Telescope, an observatory that launched beyond our atmosphere in 2009 and has shown us, in the eight ensuing years, that planets are more common than stars.

Kepler ended up being a success. But for a long time, NASA deemed it a telescope that should not exist. When Bill Borucki first brought up the idea of seeing such tiny planets passing in front of such big stars, people believed the project was impossible. "The scientific community felt it couldn't work," says Borucki. "No one had ever been able to build a photometer with the kind of precision we're talking about."

Undeterred, Borucki began working on it and first proposed a version of Kepler in 1992. "We had to get the answer," says Borucki. "Are there small planets, or not?" Back then, the mission Borucki was proposing was called Frequency of Earth-Sized Inner Planets (FRESIP). Borucki and his team proposed the mission four more times before NASA finally approved in 2000. Along the way, Tarter had a hand in helping the project move ahead. In 1995, after Borucki's proposal to NASA had once more been turned down with expressions of doubt over the ability to achieve necessary precision, David Morrison, then the director of space science at Ames, asked Tarter to review the proposed technology. When the group of experts she assembled agreed that the CCD precision would be just good enough, Tarter wrote a letter report to Morrison, urging him to find internal funding to keep the technology development going. In that same letter, she also advised Morrison to change the name from FRESIP to basically anything else, arguing that other planetary systems might not follow our template. With continued support up to the proposal's acceptance, Kepler launched in 2009, and it soon became a colossus of exoplanet studies.

In their first major announcement, in 2010, the Kepler team announced 306 potential planets. Since then, the telescope's data has revealed nearly 5,000 planet candidates and more than 2,000 confirmed ones. In our galaxy, based on what Kepler has found, about 20 percent of sun-like stars have rocky planets in their "habitable zones," where water can stay liquid. Scientist Natalie Batalha, a principal investigator on the Kepler project, doesn't just see stars anymore where she looks up at the night sky. "I see solar systems," she says.

One of Kepler's stated goals is to find terrestrial worlds. And in talk among scientists and in the headlines that splash across homepages, that often translates to excitement about finding "another Earth." That's the wrong way to think about it, says astrobiologist Margaret Turnbull, who studies the habitability of planets. "There's never going to be another Earth," she says. "We live on a planet that has a particular history . . . The *Earth* did not look like the Earth a billion years ago or two billion years ago. So no, there will never be another Earth. But that doesn't mean that life is rare or that there

aren't habitable abodes all over the galaxy. We have to liberate ourselves from the idea of these identical habitable worlds."

And when we do that, it's easier to see that planets *somewhat* like
ours are probably everywhere—and that's meaningful, good news for
SETI. "I think that helped the SETI program, the fact—the proof,
not the guess—that most stars have planets," says Borucki. "We now
know the size distribution of planets. We have information, hard
scientific program, to get those answers."

But despite the headlines, we don't actually know whether any
worlds are like Earth or whether they are habitable. "We astronomers are going into hypeland," says Tarter. "It's going to be a long
while—a disappointingly long time, I'm afraid—before we can say
for certain what a habitable planet looks like and how many of them
there are in the Milky Way."

And the distance between "habitable" and "inhabited" is a giant
leap. Our only example of a for-sure habitable world is Earth, and
our only examples for-sure inhabitation are us and our extremophile
friends. We have an *n* of 1—one solar system with one Earth with
one set of DNA-based life—and we base our predictions on that one.
"If you only have an example of one, your models are probably going
to be pretty biased and not very reliable," says Tarter.

But more examples should be forthcoming, revealing what "habitable" *really* means in this weird universe, and what might take
advantage of those conditions.

"Exoplanets are real, and now we need to make SETI real," Tarter
says now.

❖

The SETI Institute's real search for aliens has recently taken a
turn, looking for life around a type of star that—like black smokers
and nuclear cooling tanks—scientists once thought antithetical to
life. For decades, scientists thought that extraterrestrials could not
survive around most common stars in the universe—the smallest

ones, called M dwarfs, which far outnumber stars like our sun. These miniatures flared too much. Their planets would be locked, like our moon to Earth, to the star. One side would perpetually bake, while the other would stay frozen. Plus, these worlds couldn't hang on to their water. But new research has begun to suggest they might actually make okay homes, with regions of habitability and suitable wetness.

Or not. It's a debate that's ongoing in the scientific world at large and within the walls of the SETI Institute. In 2016, scientists discovered that the nearest star to Earth—a dwarf star called Proxima Centauri—has a planet at least 1.3 times Earth's mass within the habitable zone. With so many open M dwarf questions, having nature provide an exemplar so close to us is like winning the Daily Double. In the not too distant future, theory and pontification will yield to actual observation, and we will have a much better understanding of how viable these star systems are as life homes. For now, at least, they are back in the SETI game.

❖

When Tarter was at the head of the SETI portion of the SETI Institute, the Allen Telescope Array was tasked with scanning all of the exoplanetary systems that the Kepler Space Telescope and ground-based searches have found, as well as star systems in the HabCat. This habitability catalog, which astronomer Margaret Turnbull created with Tarter, identifies star systems that stay, give off the right kind of light for life, have enough heavy elements to make rocky planets, and are old enough that some advanced life form could have evolved on them.

But the ATA strategy began to change in 2012, when the CEO of the institute—Tom Pierson, who'd first taught the early SETIites to fundraise—offered Tarter a buyout. If she left, her salary would allow other members of the SETI team to keep their jobs.

She took it.

Because this was executed as a salary savings, the institute did not look for anyone external to take her job—leading the search for alien life being a hard sell with nebulous qualification requirements. So they looked inside their own office, and there they found Seth Shostak and Gerry Harp. Together, they took over the job that Tarter had done alone.

Pierson died unexpectedly two years later, after a quick bout with a bad illness, and the institute shuddered without his leadership. To fill the void, they hired David Black, briefly, as the CEO. But less than a year in, Black crashed his bicycle on one of his regular rides around the hillsides of Silicon Valley. A concussion left him in a coma, and when he came out of it, he needed more time to gain the strength needed to shoulder the stress of running the institute. In addition, board members believed they needed a new leader focused on fundraising—someone who knew the business end of a business meeting—and they pushed for a replacement who was not a science specialist but a person of the world.

In searching for Black's successor, says Tarter, "members of our board of trustees guided us away from equating our leadership with scientific aplomb and instead urged us to select a successful Silicon Valley corporate leader who understood science."

Enter Bill Diamond, a tech guy who'd worked at start-ups and Fortune 100 companies, who'd raised millions in venture capital in his past lives. The board believed he, still in the post today, had the entrepreneurial mind needed to lead an ailing institute, strapped for cash and not as connected to the Valley as it was in the Hewlett-Packard heyday, into the future.

All of this new leadership in place, the Allen Telescope Array's search strategy took a turn. Ever since Cyclops—that very first hypothetical SETI study from NASA—astronomers have said that if they looked at a million stars that were within 1,000 light-years of Earth, they would have a real statistical shot at success-fully finding ET—or saying something statistically meaningful about *not* finding ET. This in mind, Seth Shostak pointed out that *most* of the nearest stars, statistically, are M dwarfs. And since we now know that *every* star has planets and a significant fraction have rocky planets in livable places, and since M dwarfs *might* be able to host life-friendly planets, the SETI Institute should focus on these nearby neighbors. He and the leadership canceled the Kepler and HabCat search and moved on to a catalog of 70,000 red dwarfs in the neighborhood, which the ATA watches three at a time.

But the trouble is, scientists aren't *sure* planets around red dwarfs can be habitable. The dwarfs are mean, flaring out radiation that could zap nascent life, the protective atmosphere, and the water a planet was born with. The planets—even if they are warm enough for water—might be dry as the bones of aliens that couldn't survive there. And they may be "tidally locked" to their stars—one side permanently facing the suns. A habitable "belt," scientists imagine, could exist on these planets. But it's all theoretical. And it flies in the face of all past SETI logic, which goes like this: we only know of one example of life, and these are the conditions it likes, so we should look for those, because we at least know it's possible. The new strategy neglects most of the planets of that sort, around sun-like stars.

Tarter isn't pleased that the Allen Telescope Array is focusing mostly on these maybes. But she has had to give up control. It's not her institute anymore. And maybe Shostak is right.

As extremophiles showed us, humans tend to underestimate life's adaptability and overestimate its need to be "like us."

❖

Scientists like Shostak and Tarter, who look for electromagnetic signs of intelligent, tech-savvy extraterrestrials, sometimes find themselves at odds with astrobiologists: those who look for chemical signs of not-necessarily-smart, probably microscopic aliens. The separation between the two groups began back in the 1990s, soon after Congress banned NASA from funding SETI research.

At the time, Ames was NASA's official center of space life sciences, which the agency had dubbed "exobiology." Non-SETI studiers at the exobiology center began to get nervous that the reputation of the little green man and his problems on Capitol Hill would trickle down and taint perception of their own more grounded research. Wesley Huntress, the associate administrator for space science, suggested a name change—signaling a slight identity change, away from intelligent aliens and toward more traditional science—from "exobiology" to "astrobiology." On May 19, 1995, around the same time Project Phoenix started, administrator Goldin officially declared Ames NASA's center for this brand-new field.

The desire of astrobiologists to distance themselves from what's seen as a dubious endeavor has led to a kind of separation between some scientists who look for the marks of microbes and the ones who look for messages from more substantial beings. In a 2014 congressional hearing, Sara Seager of MIT, one of the world's leading exoplanet astronomers, said, "[Astrobiology is] a legitimate science now. We're not looking for aliens or searching for UFOs. We're using standard astronomy."

When Tarter hears this on the livestream, she sighs, sad and used to it. "In some sense, you can't blame Sara, right? But she certainly knows the history of what happened to SETI, and she doesn't want to get caught in that same trap," she says. "So, yes, I rankle every time Sara does that. On the other hand, I understand that from her career perspective, it's the right thing to do. It is the wrong thing scientifically."

❖

In December 2016, I had a conference call with three generations of women—Debra Fischer of EXPRES, Natalie Batalha of Kepler, and Margaret Turnbull of WFIRST, a future orbiting infrared telescope—who have dedicated their careers to searching for planets, signs of microbial life, or both, to talk about the division and occasional conflict between their fields and SETI, as well as how much the environment has changed, or hasn't, for female astronomers since Tarter sat as the lone woman in her engineering physics classes.

They all begin by saying that SETI and Tarter have been inspirational, reminding them of the big questions behind their own research. They see synergy between their fields. When Turnbull first watched *Contact*, as an intern at Harvard University, she was ready to scoff. "I was pretty sure, going into the movie, that I was going to know everything they were doing wrong because I was the smartest I'd ever been when I was a junior in college," she says, laughing. "But by the end, I forgot all about that attitude and was basically standing on my chair in the theater saying, 'That's what I'm supposed to do!'"

Not long after, in graduate school, Turnbull talked with Tarter in person. "How can somebody do their PhD with you?" she asked. She says Tarter told her that she and her colleagues were terrible graduate advisors, and she didn't recommend it. But the next summer, Turnbull went to the SETI Institute anyway to work on the HabCat. And although she doesn't do SETI now, she sees her own work—in exoplanets and astrobiology—as the best way to get close to those investigations that so inspired her in *Contact*.

The women then ask each other how many times they have each seen *Contact*, a question that is first met with *ooohs* and *aaahs*, and followed by admissions that they watch it at least once a year. No fictional science movie—not *The Martian*, or *Interstellar*, or *Arrival*—has affected them as much as Ellie Arroway's adventures and misadventures did.

But they do understand and, in some ways, sympathize with the idea that what they do is mainstream, and what inspired them about *Contact* is fringe. "Within the scientific community, there is healthy skepticism," says Fischer. "And the question is 'How do you ever get to a meaningful null result?'" Meaning, "How long and how hard do SETI scientists have to look for extraterrestrial intelligence and find nothing before they say, 'There *is* nothing. We are alone.'" And there's not a good answer, because the thing about the universe is there's always more of it to search. There are always new ways that aliens might communicate. And you could try different combinations of places and ways of looking forever and never concede.

The inability to get a null result makes a study, in the eyes of some and in some philosophies of science, unscientific. That's part of why Tarter and other SETI colleagues have tried to set limits—like looking at a million stars within 1,000 light-years—from which they can draw incremental and statistical conclusions. But then there's the whole unscientific stigma that comes from pop culture perceptions of aliens. Batalha remembers being a postdoctoral fellow at NASA's Ames Research Center, listening to a radio program about SETI and SETIites. "The host called them 'Trekkies,'" she says. "That's reflected in the politics of SETI searches."

The other women make nonlinguistic noises of assent, a practice they continue as we move on to talking about the politics of women in science.

❖

I detail for them a few things that happened to Tarter because of her womanhood: she worked alone in school because she wasn't in the boys' club; senior men took credit for her ideas; she was the only woman in many, many rooms.

Decades later, this all still sounds familiar to them. "A lot has *not* changed, and that's the problem," says Batalha, who is the only scientific coinvestigator on Kepler who is also a woman. She does not like those stats, or that feeling. But she does like when Tarter shows up at team meetings, where she's an advisor. "That helps—just having somebody in the room helps a lot," she says. *You are not alone*, the presence of another like you conveys.

On the WFIRST mission, Turnbull experiences essentially the same situation: she is the only female principal investigator. Another astronomer, Aki Roberge, works on her project and is often at meetings. "The only two women in the room: in general, that's the situation," Turnbull says. "And I've just grown so used to it over the years that I don't really think about it."

At one big meeting, she looked around at the 200 people under the same roof, 198 of whom were men. "I was thinking to myself, 'What happened here? Why aren't there—of all these teams that are responsible for all these different instruments—why is ours the only one that is run by women?'"

There are, of course, the big reasons: sexual harassment, outright discrimination, the old boys' network working its magic. Then come the slightly subtler things, like unconscious bias among hirers, leading them to prefer those who are like themselves because they happen to read the ways that white men think and act as more competent. And then we have the subtle problems that men rarely notice. Batalha recalled a time when she and a female colleague had made a point multiple times to a senior male colleague in a meeting, and in follow-up emails afterward. A week later, he came up with the same idea, and thought it was great.

This reminds Turnbull of another subtle situation that's probably familiar to most women and invisible to most men. "I could be standing in a group having a group discussion at an American Astronomical Society meeting, and unless I made a point to say something really smart, I generally felt like I would be ignored, as a woman standing there," she says. "Like, 'Maybe she should go get coffee for everybody.'"

They all sigh, thinking how different it's *not* from when Tarter started her career, and also how different even the next generation's situation isn't from their own. Batalha, whose daughter is also an astronomer, sent a picture of her with her fellow interns: she was the only woman, wearing bright colors and surrounded by guys wearing black and gray. Batalha has a nearly identical picture from one of her own internships. "I was just thinking, 'Oh my god, twenty-three years go by, and nothing's different.'" She neglected her first physics study group precisely because of these demographics. "I walked into that group once, and I felt so conspicuous I never returned," she says.

But some things are changing. When they were starting their careers, even talking about these problems, except within pink girls-only groups, was taboo. You accepted everything as the price of admission to the club of science. Now, sexual harassment has its own special sessions at American Astronomical Society meetings. There are astronomy-wide task forces to make the climate better not just for women but also for racial and sexual/gender identity minorities and those with disabilities. The changes feel slow, glacial even, but the ruling dynamics will crack soon. And maybe in the next generation, no woman (or black or gay or Muslim or transgender person) will have to be the only woman—the only anything—in the room, wondering if there is anyone else out there like them.

CHAPTER 10

SHOUTING INTO THE VOID

n 2000, Tarter's daughter, Shana, and her husband, Steve, crimped onto hope that they would match with a baby in need of parents through the Chinese adoption agency they were working with. And finally, two years after they submitted their papers, a stamp-sized picture came in the mail: this is your daughter, Li Yao, the letter said. They couldn't resolve her features, her image being so small and so distant. But they knew, finally, she was out there.

She waited for them in a foster home, where the orphanage had sent her to acclimate to life with a family, since she'd only known a room full of beds and unparented children for her whole short life. There, Steve and Shana found her, a baby swaddled in six layers of clothes. Marveling that this protoperson was theirs and that their lives would orbit a common center of gravity, they took her out into the world and unwrapped her. As they walked around town, people tugged on Shana's own clothes, telling her to put more on her baby, more between her baby and the world. "Baby cold, baby cold," they said.

❖

Two years later, Tarter went to visit Li Yao's hometown, Guilin City. It looked, to her, like a rainy Las Vegas. The city planners had replicated famous worldwide bridges to allow people to experience the alien without actually going anywhere. And all around, women carried their children strapped to their backs like turtle shells.

As Tarter traveled to the town's far northern end, the dirt roads degraded into piles of bricks, left like shrines to buildings demolished. When she reached the orphanage, she met the man who'd chosen Li Yao's name and two women who remembered caring for Li Yao. They gave Tarter a red dress to take to the girl, now two years old. Looking around at the 200 girls waiting for adoption and their own red dresses, Tarter promised to send pictures.

Li Yao, now 15, bikes farther than most adults and climbs cliffs like she belongs to a different species. She's won the state's history competition twice. Tarter speaks of these accomplishments, and those of her other family members, as proudly as if they were her own. On Twitter, she lists her progeny in her profile, and her private Gmail address is an allusion to Li Yao's name. Grandmotherhood, and motherhood, sit at the core of her identity, even if she missed many parent-teacher conferences, even if young Shana's remark about wanting to be a shopkeeper still rankles in her memory.

But as dedicated to family as Tarter is, she is also ever herself—a little flighty, distracted perhaps by big questions of the cosmos or a full schedule of conference calls. One day in March 2014, as we troll the Safeway near the Allen Telescope Array for breakfast foods (low-sugar oatmeal, bananas), decaf coffee, and a Belgian-sized bottle of locally brewed beer, she spots Mother's Day balloons floating above the line of registers.

"Shit," she says. "Is that tomorrow?"

She turns to the cashier. "Is that tomorrow?" she asks.

"It is definitely the Sunday after this one, I think," the young woman says.

Tarter whips out her iPhone to look at a calendar. She makes a note to send Shana and her stepdaughters something. One, Leslie Welch, researches visual information processing at Brown University, and the other, Jeanette Welch, plays the bass with Orchestra Iowa and in her own rock band.

All of these daughters grew up with Jill Tarter as an everyday person—a mom, a stepmom, who happened to spend a lot of time tapping on her computer and taking flights to telescopes. They were adults before she really became a public figure, a shift started with *Contact*, the movie much more than the book. But the business of minor celebrity really geared up when she was named one of *Time*'s Top 100 Most Influential People in the World, a laudatory category the magazine was just resurrecting in 2004 when they bestowed it on her.

After she had that title and those glossy photos, Tarter became a frequently requested speaker, one who could collect hefty fees that she would donate to the SETI Institute, as she and Jack also did with the money they earned from selling their plane last year.

That money made small differences, but she soon had much more substantial cash on her hands.

❖

In 2009, the phone rang in Tarter's Berkeley home, reverberating against the living room's tinted glass wall.

"Hello?" Tarter said.

"Jill," said a British-sounding male voice. "This is Chris Anderson."

Tarter knew that name, knew what it meant, and made no attempt to keep her excitement inside her skin. "Oh, Chris Anderson!" she said. "I've always wanted to give a TED Talk."

Tarter had been watching these slick, 18-minute productions—which "thought leaders" around the world give to an audience of other thought leaders who can afford the $6,000 registration fee—on her computer since they'd debuted online three years earlier, in 2006.

"This will be a very special TED Talk," Anderson's voice said over the line. "You've won the TED Prize."

Tarter hadn't even heard of the TED *Prize*—what was a TED Prize? Well, according to the TED website, it "is awarded annually to a leader with a creative, bold wish to spark global change . . . The TED Prize accelerates progress toward solving some of the world's most pressing problems." The SETI Institute's development director, Karen Randall, and board member Nathan Myhrvold, former chief technology officer at Microsoft, had nominated her without her knowledge. Randall had, concurrently, been training Tarter in the art of public persona, teaching her to keep her eyes open when speaking (she has a habit of closing them) and to tone down the jargon and speak in expansiveness.

Anderson informed her that she would receive a $100,000 award that she could use to fulfill a wish. That present came wrapped with a TED Talk and access to the organization's deep web of contacts, their expertise, and (if the winner charmed them enough) their cash.

"I know my wish!" Tarter said to Anderson, right away.

She and the institute had been struggling with the ATA, which had stood still at a battalion of 42 antennas since 2007, when it was meant to be an army of 350. The institute itself couldn't afford its scientists. While most of their researchers fund themselves through grants, the businesspeople and the people who actually perform the search for extraterrestrial smart life—not microbes or planets or water on Mars—get their salaries from the institute itself, which runs on donations. And so the heady search for alien intelligence had run up against very terrestrial constraints, the financial and political kind that have always dogged it.

"I want to build 350 telescopes," she said, putting both hands on the receiver and looking out at the bay, which had since Silicon

Valley's early days felt like a metaphor for vast, fluid possibility. "And I want to make an endowment for my scientists."

Silence came from on the other end of the line.

"That sounds like money," Anderson said. "And TED isn't really about money."

It was Tarter's turn to be silent, confused because he had just told her the prize consisted of six figures and she knew the ticket price boasted four.

"TED," Anderson continued, giving a TED-talk-esque spiel about TED, "is about getting others involved in your project to help you make it happen."

Tarter nodded and then, realizing Anderson couldn't detect it over the phone line, said, "Ah, yes."

She felt defeated, seeing the carrot on a string in front of her, having it snapped away, and being told it wasn't exactly a carrot.

"Now we have to craft your wish," Anderson continued.

"I thought that was weird," Tarter says now, "because it's *my* wish."

After emailing back and forth for a while, Anderson helped Tarter decide her heart's desire: "I wish that you would empower Earthlings everywhere to become active participants in the ultimate search for cosmic company," she said.

It's a little vague, a little hand-wave-y, and phrased to tug at your limbic system. A little, in short, like a TED Talk.

"We didn't exactly know what it was going to mean," Tarter says, "but it seemed like a good idea. The involvement of people is clearly something that I wanted to see happen, but how were we going to make that a reality?"

In her personal notebook from that year—filled mostly with calculations of antenna sensitivity and about wooing the military into telescopic collaboration—she wrote Anderson's contact details. And in stoic script next to the phone number, she wrote, "I'm a prize winner," as if she might forget.

❖

The wish would remain vague for its (standard 18-minute) TED-conference unveiling. The TED team and Tarter could figure out later how Tarter would change the world; the important thing now was to declare to an important audience that she would change (and already had changed) the world.

Tarter worked with TED's speechwriters to come up with a talk that was perfectly condensed, perfectly moving, and perfectly perspectived. The team wanted to wrap her life's work into a neat bow, as a present to all the Earthlings with streaming-quality Internet connections.

Tarter didn't really want that. Her life's work was complicated. It involved units like star-megahertz and technology like cryocoolers and calculations of how near to a star a planet had to be to have its water evaporated away. But just like TED "isn't about money," it also isn't really about depth. It's about big and important ideas presented in a way that screams, "I am a big and important idea." This stripped-down version, which told the listener how to feel instead of giving them the information and letting them decide for themselves—a mindset that undergirds scientific thought—seemed wrong to Tarter.

But in the end, Tarter drank the TED Kool-Aid, and her PowerPoint presentations have been forever altered, now favoring stock images and single-statement text over bullet points. It's helped her maintain her scientific celebrity, because it's true that the number of people who want to hear the words "star-megahertz" in their free time is small.

❖

On the day of the TED Talk, she wore a floral Japanese top and flowy black pants. She put on mascara and lipstick, feeling like she wore a mask but knowing the camera preferred the high contrast.

She stepped in front of the audience full of people she imagined might lift her up into their world of privilege, where they carefully chose where their money went because they had more than they needed, not because they had so little that they had to ration it.

"So my question," she began. "Are we alone?"

She moved across the stage and continued.

"The story of humans is the story of ideas—scientific ideas that shine light into dark corners, ideas that we embrace rationally and irrationally, ideas for which we've lived and died and killed and been killed, ideas that have vanished in history, and ideas that have been set in dogma. It's a story of nations, of ideologies, of territories, and of conflicts among them. But, every moment of human history, from the Stone Age to the Information Age, from Sumer and Babylon to the iPod and celebrity gossip, they've all been carried out—every book that you've read, every poem, every laugh, every tear—they've all happened here."

A picture of Earth popped up. Then, it flashed to a picture of the galaxy, a "you are here" arrow plopped like a T-shirt graphic at our planet's paltry location. Next, a picture of the large-scale structure of the universe, where our galaxy is one of hundreds of billions. You get the sense that you are small, worse-off than a virus squirming in the sink.

We have looked for others living their small, germy lives on other planets, but we haven't found anything. And so far the answer to Tarter's initial question—the one in her talk and in her life—is that we *are* alone.

"It's impossible to overstate the magnitude of the search that remains," Tarter continues. "All of the concerted SETI efforts, over the last forty-some years, are equivalent to scooping a single glass of water from the oceans. And no one would decide that the ocean was without fish on the basis of one glass of water."

It's a metaphor she uses often, and it works—because the ocean definitely has fish, and we know what fish are, what an ocean is.

But after this regular line, Tarter expresses for the first time the career goal that has come to dominate her late-in-life work—the one that took over once she realized she might not find life beyond Earth before her own life ends: that SETI can benefit humanity even if humans are completely alone in the cosmos, or so nearly alone we will never know we're not. And that benefit comes through a shift in worldview, or cosmosview, in which we all collectively and individually realize how similar we are—by virtue of having followed the same evolutionary path on the same planet—and that we shouldn't fight or hate or discriminate or other negative verbs that rhyme with those. We are all Earthlings, in other words. And she hopes that if people feel—really feel—that they'll start acting better.

This idea, she believes, should spread—both as an identity and as mindset that comes coupled with it. It could become "the Earthling meme," she says, spiraling into all social circles like a LOLCAT, like a virus, of which she is patient zero. But how, exactly, beyond tidy speeches, is unclear.

But back at the TED conference, the search has not ended, and neither has the TED Talk, in which Tarter stated her official wish a bit more concretely: to create a digital system in which people could actively help analyze SETI data.

"The first step would be to tap into the global brain trust," she said in her speech, "to build an environment where raw data could be stored, and where it could be accessed and manipulated, where new algorithms could be developed and old algorithms made more efficient. And this is a technically creative challenge, and it would change the perspective of people who worked on it. And then, we'd like to augment the automated search with human insight. We'd like to use the pattern recognition capability of the human eye to find faint, complex signals that our current algorithms miss."

But still, how?

❖

The TED solution began, of course, with a luncheon. Conference attendees interested in helping Tarter help Earthlings showed up for salads and iced tea. And because it was 2009, 99 percent of them wanted to build just the greatest website ever for the project.

But together, the lunchers came to the idea that the key was open-sourcing: of software, of data. They wanted to put SETI's work in the public domain so the public could participate. After all, if we are all Earthlings and the search for extraterrestrial life is so fundamental to our psyches, shouldn't we all be able to help? And so with Dell, Adobe, and a group called Galaxy Zoo, which developed a program that lets regular people classify real galaxies and spun that off into a more general citizen-science platform, they set about planning how to bring their philosophical project to the people.

Soon, a site called SETILive was born. Here, users would be able to see real-time data coming out of the Allen Telescope Array and decide whether or not any blips or bleeps came from aliens. An online place called SETIQuest became the website on which people could access and manipulate raw data and tweak the analysis software themselves, potentially finding types of signals the scientists had not thought of. The new director of open innovation, Avinash Agrawal, formerly of Sun Microsystems, was critical to the implementation of the new philosophy and the developments to back it.

But not everyone at the institute liked leaving the cathedral and entering the bazaar, letting people play around with their algorithms. They didn't want to put raw data online in real time, although that was standard operating procedure for the open-source community at large. They didn't want to clean up and document the legacy code they'd been developing since 1995 and then put it on GitHub, a repository for open-source and available software programs, for all to see.

But despite their objections, Tarter pushed them into the bazaar, and "open-source diva" Danese Cooper showed them the ropes. In 2012, at that year's TED conference, Chris Anderson unveiled SETILive as a live site, while a snowstorm pummeled the Allen Telescope Array, threatening the incoming data's nice presentation to the fancy people.

At SETILive, anyone could log on and classify signals into categories—noise or something suspicious (probably from humans but potentially from aliens)—as data poured in from the telescope. And people did do that. For a while. But then they got bored. They were stuck at desks, doing the same thing over and over again. Science like this (and really, the day-to-day of all science, no matter the discipline) is tedious. And SETI—so far, by definition—doesn't discover much, so neither do the people *doing* SETI. The user base dwindled, leaving the same few fervents appearing in the forum with their candidate signals. The institute team didn't create the technical support they needed to solve problems or innovate for the users. And eventually, in 2014, the institute shut the site down.

Today, Tarter would like to create an app that lets people swipe to classify.

"Like the Tinder of science," I say.

She asks what Tinder is.

I explain.

"So it's for sex," she says.

"Sometimes," I say.

"All social media is for sex," she muses.

"What about LinkedIn?" I counter.

"LinkedIn is definitely for sex," she counters, staring wistfully out the window of her house. "The Tinder of science. I like it."

Ultimately, she and the institute have decided it's better to leave the boring tasks to computers. Because not only do they not get bored, they are always in the process of becoming smarter and more powerful. The data that SETILive users classified came from parts of the radio wave spectrum that were regularly contaminated with human-made interference. Tarter had hoped humans could sort out those signals and figure out which spacecraft they came from, so the telescope could go back and look at those frequencies when a problematic spacecraft was below the horizon.

The telescope's real-time processing software had just been ignoring this data because it was too hard to compare, that fast, the suspicious signals to known sources of human interference, to see if they matched and could be thrown out, or if they warranted follow-up. Today, just a few years later, supercomputers at Ames are poring back over all this tossed-out data to glean from it what the humans didn't. "If they find quicker and more sophisticated algorithms for classifying this interference," says Tarter, "then they will teach the computers at the ATA to do it in near real time and open up more of the spectrum."

❖

SETILive enlisted people to help decide whether a signal is "real." But people could also help *send* a real signal into space. Distant alien citizens might, if they exist, suss out its realness and parse it for meaning. The idea that Earth should transmit messages to aliens and not just try to receive them is called active SETI or METI: messaging extraterrestrial intelligence. And in the past few years, more and more scientists have begun to take the idea seriously. The SETI Institute has initiated public discussions about the whether, how, and what of it all.

The first of these occurred in 2015 at the annual conference of the American Association for the Advancement of Science. There, Tarter had organized and was chairing a panel on which four speakers would share their contentious, contrasting views on whether active SETI might save or kill us all. And so on a balmy California February day, the spectators and press gathered in the ballroom of the San Jose Marriott, ready to hear the pros, cons, and conflicts. But as the start time approached, Tarter was noticeably not milling among them.

Five minutes before the start of the event, my phone rang, and Tarter's name appeared on the caller ID.

"My car blew up," she said.

It was unclear what this meant, beyond its consequences: The car was not in motion, and Tarter was alive but not inside the San Jose Marriott. She wouldn't make it in time for the discussion of whether or not we should broadcast messages to extraterrestrial intelligences. But the show, she said, should go on.

And so in the room where Tarter should have been, the panelists puffed themselves up for the start. These men—Seth Shostak, science fiction author David Brin, judge David Tatel, and astronomer David Grinspoon—soon took their priestly seats above the crowd.

"We're sorry our big sister couldn't be here," Shostak said, as an opener.

And then they launched in. They each had their own set of personal agendas within the meeting's official agenda, and they almost acted in two dimensions. David Brin considers himself a futurist, a forecaster ready to foretell calamities of all sorts for humanity, like the singularity, in which robots become self-aware and turn us into space-age serfs, and the possibility that if we send ET a radio message, ET will blast us with lasers and steal all our precious metals.

Seth Shostak was the fast-talking SETI—and active SETI—enthusiast, always ready with a ridiculous simile (like "life is as durable as Christmas fruitcake," from his book *Confessions of an Alien Hunter*) and spastic hand gestures. He's pro broadcasting all of Twitter out into the void, because the message would contain so much information in 140-character chunks that it could decode itself.

Judge Tatel was staid, liked order in the court, and spoke to federal precedents for large-scale decisions like whether to stream sociological commentary to the cosmos.

David Grinspoon played the part of rogue astronomer, with an upper ear piercing, soul patch, and email address that ends in @ funkyscience.net.

They talked past each other for 60 minutes and came to no conclusions.

Tarter arrived 15 minutes after the session's close, as the speakers stood near the door and the next session's audience wandered in. She promptly got in a fight with David Brin. She thinks his view is ridiculous. He thinks hers is dangerous.

❖

A 2012 document called the *Second SETI Protocol*, which has been introduced to the United Nations Committee on the Peaceful Uses of Outer Space but never officially adopted, outlines the best practices for shouting into the void—both in the case of responding to ET's call and to calling ET ourselves with no invitation. It suggests, as a start, the following three:

1.   The decision on whether or not to send a message to extraterrestrial intelligence should be made by an appropriate international body, broadly representative of Humankind.
2.   If a decision is made to send a message to extraterrestrial intelligence, it should be sent on behalf of all Humankind, rather than from individual States or groups.
3.   The content of such a message should be developed through an appropriate international process, reflecting a broad consensus.

Part of Tarter's goal with these meetings is to figure out how to come to that broad consensus and then actually have the world come to it. But no official body has adopted the protocol, unlike the protocol for how to deal with the receipt of an extraterrestrial broadcast, called the *First SETI Protocol*, developed by the International Academy of Astronautics, which has a post-detection group now headed by Paul Davies. So no one can stop your Uncle Al from setting up a transmitter in his backyard and singing Taylor Swift to Proxima Centauri b. Uncle Al, though, is not exactly what bothers Brin and others that share his opinion: it's that someone with a bigger broadcaster—an institute, say, or a country, even—could send a message saying whatever they wanted, and it would be legal.

❖

The thought leaders converge, again, on the SETI Institute on Valentine's Day, the day after the session at the AAAS, to continue their discussion and elaborate on their closely held and never-changing beliefs.

The room is set up with six long tables and a branded backdrop. A tech guy—Ly Ly—is in the process of centering the live-streaming camera straight on the institute's logo. The room populates itself slowly, never growing as dense as Tarter had hoped.

"Doug promised fifty," she says, referring to the number of crowd members and to philosopher Doug Vakoch, the institute's director of interstellar message composition. She looks around the room at the half-empty chairs.

"I've been reining in Davids Brin and Grinspoon," she informs me, quietly.

She does not want so much fighting, because too-polarized opinions always remain twain. They can't meet, let alone in the middle. And this is how she had pictured the idealized event: people with different opinions come together, they intellectualize the knots out, and then they shake hands and talk about going back to the UN. On top of that, she wanted to begin to discuss how to reach a global consensus (or at least one from a representative subset) about what a message, if sent, should actually say. This is part of her new mission, after all: to use SETI as the melting pot in which things melt into non-uniform uniformity, so the potatoes taste like the carrots and the carrots taste like the potatoes, but they retain their separate identities. It's ambitious, perhaps as unachievable as finding aliens. And the day's discussions illustrate the difficulty of deciding whether to broadcast at all, let along bringing together *the whole world's* opinions on what a broadcast should contain.

The first speaker, John Gertz, approaches the podium, and Tarter and her sheer floral skirt take a seat. She will soon believe it was a mistake not to put Gertz on her list of people in need of reining. Gertz holds the license and copyright to all things Zorro—movies, T-shirts, Z-shaped slashes people make in their curtains. He's wearing cargo hiking pants, but he speaks with the fire of someone in red leather.

"Maybe the galaxy is so silent because ET knows something we don't," he says. "That there really are planet-destroying dangers that we don't know about out there."

He goes on to describe a "berserker civilization," one composed entire of psychopaths, who might begin "quietly diverting comets in our direction" if we reveal ourselves. It sounds like a war game, like Halo: don't come out from behind that rock, or player 2 will know you're there and will shoot in your general direction until your game is over. There's no going back behind the rock.

"Earth is currently whispering its coordinates by way of electromagnetic leakage," he says, speaking about the radio and old TV broadcasts that flit past our atmosphere and out into the great beyond.

"They," he continues, gesturing in Tarter's direction, "would have us scream out our coordinates to the attention of ET intentionally."

He begins to talk about how any advanced civilization would act peaceful and perhaps pat our cute little heads, because to survive so long, violence would have bred itself out of them. That's the party line. Jill nods her head; it's something, Cold War in its logic, that she's said since the eighties. But Gertz smashes it down: he doesn't buy it, and he definitely isn't willing to bet the planet on it. He looks in the direction of the pro-broadcasters, and he holds the podium with both hands like he might pick it up and throw it out the back window and into one of the tasteful planters.

"Really?" he asks as he leans forward, fully 80 percent of his weight on his palms. "You're going to risk the entire fate of the planet just to get a conversation going?"

It may sound crazy, but danger is a possibility that can't be automatically dismissed. Because we've never met an ET, we don't know if an ET might want to stamp us out. We can say "long-lived civilizations have to be benign" all we want, but lots of predictions have been wrong, from tomorrow's weather to whether or not Donald

Trump would become president of the United States. The only thing we know for sure is that if an extraterrestrial civilization exists, its citizens were raised in a culture, and possibly with a moral system, totally different from ours. They might not even have eyes, let alone diplomacy.

"Scientists who I otherwise consider reasonable throw logic out the window," Gertz closes. "With METI, we're playing Russian roulette, and we don't know how many bullets are in the chamber."

Broadcasting should be illegal, he concludes. Future generations should decide if they are ready, rather than having us decide their fate for them when armies of laser blasters show up 1,000 years from now after a 20-light-year trip.

Tarter smiles and shakes her head in the front row, scribbling notes in the way I used to admire in adults, her hand shifting across the page and making that *scritch scritch* of tiny friction you can hear from a few feet away.

The next speaker saunters to the podium. He's a lawyer, crisp and ultra-American in appearance: Adam Korbitz.

"We do not have enough evidence to conclude active SETI is inherently risky," he says.

But people who advocate against it follow the precautionary principle, even if they don't know that legal jargon: protect against potential harms, even if causal chains are unclear and we don't know if these harms are founded.

"The weak version," he says, "is better safe than sorry. Do not require unambiguous evidence of harm before taking protective action."

It's meant to protect, like a gun in the dresser drawer. But like a gun in the dresser drawer, it may do the opposite. Being too cautious could cause us to miss out on vital cosmic information, like what the hell is up with quantum mechanics and how to stop our planet from becoming a climate-changed hellhole.

"It would seem colonization, et cetera, would be worse than losing out on a little help from our friends," Korbitz said, "except we face many challenges to the long-term survival, which may depend on joining the galactic club."

Speculating about the dangers, in other words, is no less speculative than speculating about the benefits.

Tarter is back to nodding, and Brin and Gertz are shaking their heads. It's a room full of cranial sloshing.

In 2016, believing that the SETI Institute was too split on the issue of broadcasting, Douglas Vakoch, who had been the institute's director of message composition, split off to form his own organization, METI International. By 2018, they plan to be broadcasting, board members and naysayers be damned.

❖

So if people are going to broadcast, how are we (or they) to decide what to say? The truth is, we have sent a few pings out into the universe already. In 1974, astronomers blasted a powerful radar signal from the Arecibo telescope, whose dish amplified it 10 million times. It contained some remedial math, the basics of DNA, a dot-matrix picture of a human, and a map of the solar system, among other things.

Three years later, Carl Sagan, Ann Druyan, Timothy Ferris, and Jon Lomberg created a physical proclamation for aliens to find: the famous Voyager golden records. The twin *Voyager* spacecraft launched in 1977, on courses to pass by the picturesque Jupiter, Saturn, Uranus, and Neptune, snapping the kind of up-close tourist photos that would have made them Instagram famous. But then, they just continued voyaging—always voyaging—out beyond the solar system. Inside their spidery, synthetic bodies, though, live two gold discs onto which are etched 54 sound clips, 90 minutes of song, 116 images, and greetings in 55 languages. Inhabitants of a distant star system, or perhaps aliens themselves voyaging through the cosmos, might someday find our dinky-looking spacecraft, pluck out the record, follow the instructions, and hear a dog bark or a rocket lift off; they might look at *World Book*'s anatomy drawings or snowfall on sequoias. They could listen to Ann Druyan's heartbeat. The four attempted to take the experience of Earth—all Earth—and condense it down onto this compact disk. And they did a pretty good job. But they were just four white people in a room, making decisions for the whole world.

Don't feel too slighted, though: even if extraterrestrials exist, they arc unlikely to find those spacecraft or our electromagnetic missives. Leakage from *I Love Lucy* is 60 light-years away, but those broadcasts will have quieted so much—their volume turned down by the long journey through space—that an antenna many times bigger than any humans currently have would be required to pick them up. Our purposeful broadcasts, like that from Arecibo and two sent from Yevpatoria in Russia, were of short duration: an inhabited planet would have had to be tilting its electromagnetic ears in our direction at the exact right time, and what are the chances of that?

Slim.

And the physical artifacts, like the Voyager record—their chances are much, much slimmer. No one is going to find them unless they have infinitely good luck or whatever the cosmic equivalent of a metal detector is.

But what the scientists sweating and pointing fingers in that Silicon-Valley conference room are discussing is a bit different: sending a strong, purposeful broadcast, like a military-grade radar blast—perhaps constantly, perhaps for years, maybe thousands. Aliens wouldn't have to be trying particularly hard at a particular time to stumble upon it.

And what if they did find it? What would they learn about us? Well, that depends on what we tell them.

❖

Figuring out *what* to broadcast, especially if our missive is supposed to somehow represent all of humanity, is a task for patient geniuses, deep-learning algorithms, or, if we listen to Shostak, Twitter. In our past, the Voyager message was the most serious attempt to represent civilization. But it was hardly comprehensive. "I thought we had lied through our teeth. It was a really good effort, but it was unrepresentative of the Earth—no poverty, no disease, no war, no fame," says Tarter. "Put your best foot forward."

On top of that was its insularity. What about the *rest* of the Earthlings? A new project called the One Earth Message hopes to be an update, and an improvement. Led by artist Jon Lomberg, who worked on the Voyager message as well as Sagan's *Cosmos* television series, it aims to capture the crowd. Anyone with an Internet connection can send in pictures, video, or audio that they feel someone capture our experience on Earth. Lomberg then hopes to broadcast the message to NASA's *New Horizons* spacecraft, which flew past Pluto in 2015 and is headed out into the wider universe. Once *New Horizons* has completed its extended science mission, flying by the Kuiper Belt object 2014 MU69 in 2019 and sending its scientific data back to Earth, scientists could clear its memory to make room for a new message: a digital and more representative version of Sagan's analog, etched record that would fly with *New Horizons* until space radiation begins to deteriorate its bits in many millions of years. The One Earth Message, which NASA hasn't yet approved but for which it has expressed approval, will certainly be more inclusive than the Voyager record, which four people in a room put together. But still, just half of the world's population has Internet access. And the media that ultimately make it on to the craft will be chosen by some kind of voting or voting-and-algorithm combination. Neither method is neutral. A dominant group will have a dominant say, and someone has to write the algorithm, which means it will contain their or their culture's biases.

But the One Earth message is a start, and an attempt to out find out what we—as individuals and as a collective—value, how we think of ourselves. Maybe as important, we may find out how our self-images differ, which could jolt us, like finding out that a best friend has an opposite interpretation of a poem that we've read during every hard time in our lives. Or like finding out that your best friend saw the Dress as blue and black, not white and gold. It's a good way to work together and sing kumbayah and think about the long-term future of humanity and its place in the cosmos. But it's probably not going to be truly representative, and it's probably not going attract the attention of beings beyond Earth. The most important audience for the message is ourselves.

And maybe that's good, because scientists haven't historically had success gathering input from the crowd. From the crowd came no wisdom. A few years ago, for instance, the SETI Institute tried the Earth Speaks program, in which people sent in the words and images in answer to the question, "What would you say to an extraterrestrial civilization?"

Some people drew My Little Ponies; others wrote "Don't eat me."

The Deep Space Network sent 130,000 Craigslist messages to aliens. For $99, you could even record a five-minute voice message. TalkToAliens.com charged people $3.99 per minute to record their greeting. In the 2008 "A Message from Earth," Russian astronomer Alexander Zaitsev used the social site Bebo to collect 501 textual and visual messages for solar system Gliese 581, including side-by-side photos of George W. Bush and Barack Obama: Good, evil, get it?

Hint: They won't. Without a Rosetta Stone of sorts, all our collective shoutings and scrawlings are meaningless, even if an alien *does* intercept them. But the One Earth message has a veteran at its helm and NASA's support, and is dedicated to inclusion, even if the chances of interception stay slim. And METI International has powerful telescopes and the world's foremost expert in how we could make extraterrestrials understand us.

❖

But now there's another big player in the interstellar message game—a kind of competitor—that also wants to create such a message. This initiative is called Breakthrough Message. It all started with Yuri Milner, a Russian billionaire who made his money in Internet entrepreneurship. But Milner had begun his career intending to be a physicist. "And when he looked at a list of the most famous people in the world, and none of them were scientists," Pete Worden said at the 100-Year Starship conference in 2015. "He wanted to change this."

And so he founded the Breakthrough Foundation, which dispenses the Breakthrough Prizes, which Worden called, in the same talk, the Academy Awards of science. The award ceremony itself, held in a hangar at Ames Research Center where SETI got its start, is the only black-tie event in hoodie-saturated Silicon Valley. Worden was the director, and so the landlord, of Ames at the time the awards began, and that job brought him face-to-face with Milner. They got to talking about SETI, and Milner, as people with money decide to do sometimes, wanted to start and sponsor a program.

"They asked if I knew anyone who could run it, maybe someone who had run a NASA center," said Worden, metaphorically winking. "So I put in my paperwork."

He is now the chair of the Breakthrough Prize Foundation and represents Breakthrough Message, which aims to encourage discussion about whether to send something to extraterrestrials—and what and how. The organization plans to soon open a competition—with a $1 million prize—for the best and most representative digital messages. But the dates and details are still TBD, and the team promises "not to transmit any message until there has been a global debate at high levels of science and politics on the risks and rewards of contacting advanced civilizations," according to their website. Good luck.

The Breakthrough Foundation also runs Breakthrough Starshot, which is investigating how to send a tiny spacecraft to the close-ish star system Alpha Centauri, and another project, called Breakthrough Listen, which is a traditional SETI program headed up at the University of California, Berkeley.

Breakthrough Listen's existence reads a bit competitively, the new manifestation of the Berkeley-SETI Institute divorce that happened during the recession. Everyone is nice to each other: Dan Werthimer, who once worked with Jill on collaborations between the SETI Institute and Berkeley, now leads the Listen program. And Tarter sometimes attends their meetings and sits on their advisory committee. There's room on Earth for two SETI programs; after all, it's not like there's only one Milky Way Galaxy Research Center on the planet. But there's a bit of bad blood there: Breakthrough began without Tarter and the ATA—and without their advance knowledge, which, in a small community, felt like a slight.

Breakthrough Listen came to be when Berkeley astronomer Geoff Marcy (now as famous for sexually harassing students, according to a 2015 Title IX investigation that led to his resignation, as he is for discovering many, many exoplanets) gave a talk about said planets at Milner's mansion in Silicon Valley. The mansion, not coincidentally, once belonged to Barney Oliver, SETI's former benevolent warlord. Marcy's talk extolled the virtues and successes of exoplanet science.

"At the very end, he showed a few slides about SETI, what he was doing with SETI, what I was doing with SETI," says Werthimer.

That caught Milner's attention, and he asked if there was anything he could do for the field. The answer, pretty much always, is money. So soon, Milner promised the team $100 million total over 10 years of support.

"I spent my whole life trying to raise funds," says Werthimer. It was never this easy, or this substantial.

Considering how most of Tarter's career was spent worrying about the checkbook, and how many projects limped or faltered because of financial concerns, that rings true, and stings.

And Tarter heard about the project first not from the Breakthrough people but from Tony Beasley, the head of the National Radio Astronomy Observatory, which then ran the Green Bank Telescope. Marcy asked Beasley if Breakthrough Listen could pay for time on that telescope. Beasley, surprised by the request, called Tarter to ask if Marcy was "acting on his own."

"Look," Tarter told him, "I can't tell you what Geoff is doing because he hasn't told me."

And Marcy did not, in fact, tell her till the day the project went public, and neither did anyone else.

When, after the Title IX revelations, Marcy was forced to step down from his positions at Berkeley and with Breakthrough, a young guy named Andrew Siemion took over. He and Werthimer now lead the charge. Their project is flashy, with its many millions and its use of big telescopes: Green Bank, Parkes, and, soon, China's new 500-meter FAST antenna. Stephen Hawking and Mark Zuckerberg sit on the initiative's board.

But more than a year into the project, the team has hardly analyzed any of the many terabytes of data they have collected. Their philosophy—get the data—is different from the SETI Institute's, which is, "Get the data and dig into it right away, so that if you find an intelligent shout from the void, you follow up and confirm."

Breakthrough, though, has focused on *getting* the data, starting observations at new telescopes, and figuring out what to do with their abundance of bytes later. And they do not use a back-up telescope, like Tarter and the SETI Institute did during Project Phoenix, to look at the same spot in the sky as the main telescope, confirming it sees the same thing. Tarter believes these differences between their program and hers are mistakes. But Werthimer feels confident they'll eventually get their data analysis under control and can do confirmation differently.

"I've observed for decades, deliberately doing something else because I thought it was a better idea," says Tarter. "But it doesn't mean that I'm right and they're wrong."

Breakthrough Listen has also invested resources in a different kind of SETI: optical SETI, which looks for bright laser beacons that pulse fast. First suggested by Nobel Prize winner Charles Townes in the 1960s, this kind of search has only recently become financially and technologically feasible. Other groups, including the SETI Institute itself, a Harvard group led by Paul Horowitz, and a UC

San Diego team helmed by astronomer Shelley Wright, have also dipped into this new kind of observation. But moving from radio-only to radio and optical is likely only the beginning of the search expansion. "People say, 'Fifty years, that's such a long time, and you haven't found anything,'" says Tarter. "And then you think of all the things we're not looking for, haven't had the capability to look for, don't even know to look for. We reserve the right to get smarter."

❖

The third Breakthrough SETI program, Starshot, plans to figure out how to send a probe to the Alpha Centauri star system. The leaders of Starshot are pioneering "lightsail" technology: they plan to launch a swarm of tiny spacecraft with a photon-catching sail and then send pulses of laser light—essentially just energy—to the sail, pushing the craft to 100 million miles an hour. At that speed, it could reach the planet Proxima Centauri b in just 20 years and then relay images back to Earth of what that world is like. Such a spacecraft, they estimate, could be ready within a generation. The mission was already in place as an idea when astronomers from a team calling themselves Pale Red Dot, a reference to Carl Sagan's statement that Earth in cosmic context is just a pale blue dot, found an Earth-sized planet orbiting the star Proxima Centauri.

Tarter, meanwhile, has been part of another deep-space exploration program—the 100-Year Starship program—that, maybe someday, will involve humans.

❖

In September 2014, music blasts from the speakers of the George R. Brown Convention Center in Houston: the Beastie Boys' "Intergalactic," Justin Timberlake's "Spaceship Coupe," Nicki Minaj's "Starships." The clock reads 20 minutes after 9 A.M., 20 minutes after the 100-Year Starship symposium was supposed to start. Tables with galaxy-speckled tablecloths constellate the room, which continues to fill with rocket scientists, science fiction authors, psychologists, microbiologists, self-proclaimed futurists, and curious spectators willing to front a registration fee. Despite their desire to embark on an interstellar journey sometime in the next century—the stated goal of 100-Year Starship collaboration—they seem in no hurry to begin this year's discussions. Across the table from Tarter, a man dressed in steampunk clothes asks what the Wi-Fi password is.

The Wi-Fi password is *extraordinary*.

Tarter is reading and deleting emails on her iPad. She notes in a voice directed at her screen that attendance is lower at this conference than it has been in the past few years. 100YSS, as the group calls themselves, has been meeting since 2011, when they won a contract from the mysterious Defense Advanced Research Projects Agency. Mae Jemison, the first female astronaut of color, heads the foundation.

If you Google "DARPA + creepy," you'll get about as many results as if you Google "Jill Tarter." DARPA handles the development of "emerging technologies" that might somehow be useful to the military. The military can appropriate and make use of pretty much anything, and DARPA invests in the "frontiers" of science and technology with the hope that pretty much anything advanced can advance spying, defending, and killing. Ironically, the agency began as a response to the Soviets' launch of Sputnik in 1957, to ensure that the US would never again be behind. Or, as they put it, "to prevent technological surprise like the launch of Sputnik." They've expanded that from defensive to offensive, hoping not only to forestall but also to create technological surprise, leaping out from under-the-radar planes with invisibility cloaks and laser-based weaponry. They've made robots that cross uncomfortably into the Uncanny Valley; they've developed satellites that see in Soviet windows; they have a project called Combat Zones That See, in which a web of connected cameras tracks everything in a city that moves; they're looking in your brain to see how it fires when you say a specific word, and then using that knowledge to implant words in your head. Or to put it more simply, telepathy. And in 2010, DARPA teamed up with NASA's Ames Research Center to fund the organization with the best business plan to foster research into interstellar travel, hold up for 100 years, and (just maybe) end with a one-way trip to some godforsaken star system. Despite that lofty goal, the Starship program is mostly still conferences, and the initial grant—which mostly covered those conferences—is gone. Still, the organization remains in existence, and Jemison hopes to continue to use it and the community of thinkers and doers she has assembled to propel humans to the stars.

Tarter is interested in the starship itself, and the journey, and other star systems, and generally being starry-eyed. But the conference is, in general, a bit starry-eyed. At the same time that an outsider might feel inspired, walking in to a meeting that feels so proactive and forward-watching, she might also find herself a little wary of the motivational-poster talk going on. Your sincere brain wants to believe; your cynical brain wants to scoff and ask if you accidentally stepped into a cosmic Comic-Con.

Tarter keeps a lump of skepticism in her throat, at the ability of humans to keep organized and not let daily friction slow the project to a stop (especially once the founders die and somebody's children have to take over). But also at the possibility of interstellar travel itself. "I can envision very smart, small probes being sent and surviving," she says. "But I don't know about with big, wet technology—biological space travel. But because I don't know, that's precisely why I signed on to the advisory committee for the 100-Year Starship study. I take Arthur Clarke's second law—that the only way to know what's possible is to venture a bit into the impossible—quite seriously." (That phrase may, in fact, be on one of the posters outside the conference room.)

But, she continues, "as we tackle these challenges and begin to succeed and find solutions, that's going to be benefit life on Earth."

And that's the reason DARPA funded 100YSS—a reason that's also right in its slogan: Pathway to the stars; footprints on Earth.

People are more motivated to build, say, better batteries for spaceships than they are for soldiers, but the batteries can, indeed, power both vessels. A self-sustaining life-support system could save (to give the archetypal example) starving children in Africa, isolate the source of some blood-curdling new epidemic, or keep people alive on a spaceship. And DARPA might care some about that, or the fact that such a system would make an excellent field hospital. As their original goal states, their real aim is to acquire technology they can use to defend and surprise "the enemy."

Like with radio astronomy and SETI, there has always been an uncomfortable relationship between space exploration and the military. In the early to mid 20th century, many great American physicists got their starts working on the Manhattan atomic bomb project. We only went to the moon so the Soviets wouldn't put boots on its ground before us. NASA has top-secret security clearance departments, housed in buildings with fake, bricked-over windows.

Science doesn't get very much money; F57s do. Any time scientists can do work for war departments, science benefits, monetarily. But the relationship does seem a bit uncomfortable given that DARPA began in an effort to make a giant leap over Soviet space travel, while a mission to a different solar system would almost have to be an international project. No single country—not even the United States or China—has the money or resources to go it alone. And, besides, once the colonists left, other countries probably wouldn't be too pleased that English speakers were the only ambassadors to Gliese-166b. It wouldn't be very *Earthling* of us.

Finally, at the conference, Jemison walks toward the front of the room. She puts her fingers on Tarter's shoulder as she passes, and their cape-like cardigans brush against each other.

"Hi," they say.

Jemison continues up to the podium to begin the conference and introduce the first session: about a NASA brainstorming retreat, where rocket scientists got together to discuss the feasibility of long-term interstellar travel. It is a report on a brainstorm at a conference that is also a brainstorm. Throughout the whole panel, Tarter seems to be composing emails and paying only scant attention. But at the end of this panel—and nearly every other panel—she raises her hand and asks a straight-on question.

"And what actual, tangible results has the agency gotten from this meeting?" she asks the NASA scientist.

❖

She probably asks others this question as she asks it of herself all the time. She has worked for more than four decades on a project that some believe to be silly and others to be futile, wasteful, hubristic, or any of a whole host of other adjectives. And she has not been able to find any evidence to answer the question that she set out to answer: are we alone?

Those words rattle around in her head. And after forty-plus years of study, she's less certain about the answer than she was when she was a kid standing on a beach, looking up at a star and imagining someone looking back, catching her eye across the cosmos. After all, that catch hasn't come.

Tarter sometimes expresses disappointment—in herself, in the endeavor—for this inconclusiveness. At the end of the phone call with the three generations of exoplanet astronomers, I ask if there's anything I missed—about Tarter, about planets, about the universe— that they would like to say. Batalha jumps in to express solidarity with the non-conclusion of the conclusion of Tarter's career: their fields—SETI, astrobiology, exoplanet science—require generations of work. All of science does, really. Big discoveries are rare, coming decades or centuries after people start wondering and doing the work that scaffolds them, shores them up, sets them up to succeed. But without that initial wondering, and those first small steps, no one would make giant leaps at all. "Jill has had this really luminous career doing SETI," says Batalha. "But at the end of the day, she retired and hadn't found anything. And I'm guessing that might be my fate as well, in terms of finding [microbial] life. I might live to see that day, or I might not."

To be an astronomer at all is to be zen about that: about cosmic time and about how you are a cog in the big machine of science, whose gears began turning long before you and will continue to turn long after you. Sometimes those gears grind to a result because of your cog, and sometimes your cog is just there to keep the gears going.

All astronomers have days when they're good at being zen, and days when they feel hopeless about and powerless before the uncaring bigness and seeming incomprehensibility of the universe. Tarter has had more of the latter recently.

Batalha recalls a Kepler meeting, as the project's prime data collection time was ending, in 2012. She was sitting next to Tarter, who, at a certain point, looked down at the table and near-whispered, to no one but herself, "We didn't find anything."

Batalha turned her head to look at Tarter, struck by the depth of emotion. "That feeling—it was just so tangible," she says. "She announced her retirement two weeks later. Clearly, she knew that she was on the verge of retiring. She was expressing that feeling of all those years of work not realizing that goal."

She pauses, perhaps to feel that feeling again for a second, taking it on like Jonas in *The Giver*. And then she continues, reaching back toward the peace of a cosmic perspective, where an individual and her lifetime are just blips. "But that's how science works," Batalha says. "It is what it is."

Turnbull chimes in, on the call, to say she had a similar experience at Tarter's retirement party. She was catching up with Tarter's long-time administrative assistant and unflagging companion, Chris Neller. Jill, Neller told her, feels like a total failure and that everything she did was for nothing.

"And I just thought, 'How could she really?'" says Turnbull. "Because it never even occurred to me, because of the enormous impact that Jill has had—and that thanks to her we are all continuing to have—on the world, just simply by existing and being in the scientific community and pushing the frontiers in every direction."

And then Turnbull points out something about Tarter that's not true of most scientists in most fields: "She's not replaceable," she says. "Many things would not be the way they are if not for the work that she did, and the career she had."

For decades, Tarter was *the one* who championed the search, through its death, into its resurrection, as it rode the roller coaster of its modern incarnation. She had the support of her colleagues, but she was the keystone in their arch: without her, it would have collapsed. Sure, someone would have eventually picked up the rubble and rebuilt. But because of Tarter, no one had to.

Her perseverance reminds me of one scene in *Contact*, when Ellie Arroway's nemesis and former advisor David Drumlin shows up at the Arecibo telescope. He's shutting her SETI program down, he says, the surface of the radio dish looming over him.

"I know you can't see it now, but I'm doing you a favor," he says. He's protecting her from her own inclinations. "You're far too promising a scientist to be wasting your gifts on this nonsense," he says.

Arroway's nostrils go wide with indignation. "Look, I don't consider what could potentially be the most important discovery of the human race nonsense, okay?" she says. "There's 400 billion stars—"

"And only two probabilities," Drumlin interrupts. "One, there *is* intelligent life out there, but it's so far away you'll never contact it in your lifetime, and two—*two*—there's nothing out there but noble gases and carbon compounds, and you're wasting your time. In the meantime, you won't be published, you won't be taken seriously, and your career will be over before it's begun."

"So what?" says Arroway. "It's *my* life!"

So what? Tarter's actions have conveyed. *It's my life.*

❖

In Tarter's nonfictional life, she was published. She was, and still is, taken seriously. Her career was not over until she decided it was. She may not have unearthed an extraterrestrial civilization or made the most important discovery of the human race, as Arroway did. But she helped SETI survive, and she helped others think bigger, be better, and stay stubborn. And if humans ever do find out we're not alone, we will all be tipping our hats—or whatever people wear on their heads in 2050 or 2500 or whenever the metaphorical phone may ring—to Tarter, for keeping the search alive, for keeping that unanswered question lodged in our brain folds.

No, we didn't find anything. But the attempts mean something.

❖

In Tarter's office is, among her many treasures, a piece of collage art. It depicts an ATA antenna in its Hat Creek habitat. The artist, Inger Friis, created the collage in her late 90s, having recently learned that form after decades of Chinese finger painting. Spinal cancer had caused problems that prevented standing at an easel perfecting hues for hours on end as she had in her youth, so she learned to work sitting down with just fingers and fingernails. And after she mastered that, and before she turned 100, she decided to try something new. For her, it was never too late.

Tarter met Friis after her husband, Harald—a famous radio engineer—passed away and donated part of his estate to the SETI Institute. While talking to Friis about that transaction, Tarter learned that all of Friis's finger art was hiding under her bed in a Palo Alto retirement home.

Tarter hated the thought of a life's work languishing in dusty darkness. And so she asked Friis if they could exhibit the work at the institute. Not long after the initial call, Friis's art stood on easels in those halls, next to posters about Martian craters and dolphin communication.

Tarter and Friis became close, talking long after the paintings left the institute walls and found new homes. Their discussions often drifted toward aging and dying—because Friis was probably dying, because Tarter has always wanted to know what her own future will look like. "How does it feel?" Tarter asked.

"You lose a piece of yourself every day," Friis told her. "That's the process."

But works of art—like scientific contributions, and social ones—leave permanent marks on the world's surface. Fossils, big or little, that indicate to Earth's future inhabitants *We were here.*

A TIMELINE OF SETI

1959   Guiseppe Cocconi and Philip Morrison publish the first scholarly SETI paper, in the journal Nature.

1960   Frank Drake performs the first SETI search—called Project Ozma—with a radio telescope in Green Bank, WV.

1961   The National Academy of Sciences convenes the Order of the Dolphin meeting in Green Bank, WV, using the Drake Equation as an agenda.

1970   NASA's Ames Research Center undertakes the Project Cyclops study.

1971   Project Cyclops is published, and the first joint US/USSR SETI meeting happens in Burykan, Armenia.

1974   NASA establishes the Office of Interstellar Communication under John Billingham.

1977   NASA publishes its Workshops on Interstellar Communication.

1979   Senator William Proxmire gives his Golden Fleece Award to SETI.

1981   Senator Proxmire cancels funding for NASA's nascent SETI program.

1982   The federal government reinstates SETI's funding, and Suitcase SETI observations begin.

1989   Scientists create the "post-detection protocol": "Declaration of Principles Concerning Activities Following the Detection of Extraterrestrial Intelligence."

1993   Senator Bryan terminates NASA SETI funding. Scientists discover the first planets beyond our solar system, around a pulsar.

1995   Project Phoenix observations begin at Parkes and Mopra Observatories in Australia. Astronomers discover planet Peg 51b, the first around a Sun-like star.

1996   Project Phoenix observations begin in Green Bank, WV, and Woodbury, GA.

1998   Project Phoenix observations begin at Arecibo and Jodrell Bank. Astronomers perform the first optical SETI observations at Harvard and Berkeley.

2000   Paul Allen funds technology development for the Allen Telescope Array.

2003   Paul Allen funds the first phase of construction for the Allen Telescope Aray.

2007   Engineers install the 42nd and final dish at the Allen Telescope Array, which is then dedicated.

2009   Radio science and SETI observations begin full-time at the Allen Telescope Array. Tarter wins the TED Prize.

2011   The University of California, Berkeley, ends its partnership in the Allen Telescope Array, and SRI becomes the operating partner with the SETI Institute. The Kepler Space Telescope releases its first batch of exoplanet data.

FURTHER READING

Billingham, John. *Social Implications of the Detection of an Extraterres-trial Civilization: A Report of the Workshops on the Cultural Aspects of SETI.* Mountain View, CA: SETI, 1999.

Billings, Lee. *Five Billion Years of Solitude: The Search for Life among the Stars.* New York: Current, 2013.

Cameron, A. G. W. *Interstellar Communication.* New York, Amsterdam: W. A. Benjamin, 1963.

Cultures beyond Earth: The Role of Anthropology in Outer Space. New York: Vintage, 1977.

Davies, P. C. W. *The Eerie Silence: Renewing Our Search for Alien Intel-ligence.* Boston: Houghton Mifflin Harcourt, 2010.

Dick, Steven J. *The Impact of Discovering Life beyond Earth.* Cam-bridge: Cambridge University Press, 2016.

———. *Life on Other Worlds: The 20th-Century Extraterrestrial Life Debate.* Cambridge: Cambridge University Press, 2001.

————. *Many Worlds: The New Universe, Extraterrestrial Life, and the Theological Implications*. West Conshohocken, PA: Templeton Press, 2001.

Dick, Steven J., and Mark Lupisella. *Cosmos and Culture: Cultural Evolution in a Cosmic Context*. Washington, DC: National Aeronautics and Space Administration, Office of External Relations, History Division, 2009.

Dick, Steven J., and James Edgar Strick. *The Living Universe: NASA and the Development of Astrobiology*. New Brunswick, NJ: Rutgers University Press, 2004.

Ekers, Ronald D., and Philip Morrison. *SETI 2020: A Roadmap for the Search for Extraterrestrial Intelligence*. Mountain View, CA: SETI, 2002.

Fields-Meyer, Thomas. "The Searchers." *People* 52, no. 13 (October 4, 1999).

Finney, B. "SETI, Consilience, and the Unity of Knowledge." *Bioastronomy '99: A New Era in Bioastronomy*, Conference Series 213. San Francisco: Astronomical Society of the Pacific, 2000.

Garber, S. J. "Searching for Good Science: The Cancellation of NASA's SETI Program." *Journal of the British Interplanetary Society* 52, no. 1 (1999): 3–12.

Grinspoon, David. *Earth in Human Hands: Shaping Our Planet's Future*. New York: Grand Central Publishing, 2016.

Gulkis, S., E. T. Olsen, and J. C. Tarter. "A Bimodal Search Strategy for SETI," in *Strategies for the Search for Life in the Universe*, ed. M. D. Papagiannis. Dordrecht, Holland: D. Reidel Publishing Co., 1981.

Harrison, Albert A. "The Relative Stability of Belligerent and Peaceful Societies: Implications for SETI." *Acta Astronautica* 46, no. 10–12 (2000): 707–712.

————. *After Contact*. New York: Basic, 2002.

Impey, Chris. *The Living Cosmos: Our Search for Life in the Universe*. New York: Random House, 2007.

Irion, Robert. "What Proxmire's Golden Fleece Did for—and to—Science." *Scientist* 2, no. 23 (December 12, 1988).

Kaufman, Marc. *First Contact: Scientific Breakthroughs in the Hunt for Life beyond Earth.* New York: Simon and Schuster, 2011.

Miller, Ben. *The Aliens Are Coming! The Extraordinary Science behind Our Search for Life in the Universe.* New York: Experiment, 2016.

Morrison, Philip, John Billingham, and John Wolfe. *The Search for Extraterrestrial Intelligence.* Washington, DC: National Aeronautics and Space Administration, 2011.

Nagel, Thomas. "What Is It Like to Be a Bat?" *Philosophical Review* 83, no. 4 (1974): 435.

Oliver, Bernard. *Project Cyclops.* Washington, DC: National Aeronautics and Space Administration, 1973.

Poundstone, William. *Carl Sagan: A Life in the Cosmos.* New York: Henry Holt, 1999.

Sagan, Carl. *Contact: A Novel.* New York: Simon and Schuster, 1985.

Sagan, Carl, and I. S. Shklovski. *Intelligent Life in the Universe.* New York: Doubleday, 1980.

Scharf, Caleb. *The Copernicus Complex.* New York: Scientific American/Farrar, Straus and Giroux, 2014.

Schuch, Paul H. *Searching for Extraterrestrial Intelligence: SETI Past, Present, and Future.* New York: Springer, 2011.

Schwartz, Peter. *The Art of the Long View.* New York: Doubleday/Currency, 1991.

Shostak, G. Seth. *Confessions of an Alien Hunter: A Scientist's Search for Extraterrestrial Intelligence.* Washington, DC: National Geographic, 2009.

Sobel, Dava. *Is Anyone Out There? The Search for Extraterrestrial Intelligence.* New York: Pocket, 1997.

———. *The Glass Universe: How the Ladies of the Harvard Observatory Took the Measure of the Stars.* New York: Viking, 2016.

Swift, David W. *SETI Pioneers: Scientists Talk about Their Search for Extraterrestrial Intelligence.* Tucson: University of Arizona, 1990.

Tarter, Jill. "SETI Observations Worldwide," in *The Search for Extraterrestrial Intelligence,* ed. K. I. Kellerman and G. A. Seielstad. Green Bank, WV: NRAO, 1985.

Tarter, Jill, and M. Michaud. "SETI Post-detection Protocol." 37th and 38th Conferences of the IAF. International Astronomical Federation, 1987.

Tough, Allen. *When SETI Succeeds: The Impact of High-Information Contact*. Bellevue, WA: Foundation for the Future, 2000.

Vakoch, Douglas A. *Archaeology, Anthropology, and Interstellar Communication*. Washington, DC: National Aeronautics and Space Administration, Office of Communications, History Program Office, 2014.

———. *Psychology of Space Exploration: Contemporary Research in Historical Perspective*. Washington, DC: National Aeronautics and Space Administration, 2011.

White, Frank. *The SETI Factor: How the Search for Extraterrestrial Intelligence Is Changing Our View of the Universe and Ourselves*. New York: Walker, 1990.

ACKNOWLEDGMENTS

'd like to thank my family—Darla, Ron, Rachel, and Rebekah—for everything; my grandparents, Eleanor and Floyd Kinney, for indulging me in big conversations when I was little; Brooke Napier, for unconditional support and strategic frozen yogurt; Ann Martin and Joshua Kinne, for their insights and deep friendship; everyone in Green Bank, for being astronomical badasses; Tasha Eichenseher, Breanna Draxler, Lisa Raffensperger, and Siri Carpenter, for helping me shape the book when it was a proto-idea; Miriam Kramer, for being my space friend; Christopher Kempf and Matthew Grice, for being my writing buddies; Dana Koster and Justin Souza, for the cheerleading and the cabin; Amber Dermont, for teaching me that I wanted to write stories; Chris Depree and Amy Lovell, for teaching me astronomy; Kim Dahl, my sixth-grade science teacher, for showing me what inquiry means; Jane Cooper, my high school English teacher, for all the sentence diagrams and smart jokes; the Cornell MFA program, for giving me a toolbox; Katie Palmer and Adam Rogers, for giving me a shot; my secret science-writer chat groups, for existing; the SETI Institute interns, for letting me tag along; Chris Neller, for letting me in to the SETI Institute and keeping immaculate records; Karen Randall, for the jumpstart; Vera Buescher, for keeping track of all the politics; Carl Sagan, for *Contact*; the Wellstone Center in the Redwoods, Melody Keen Haller, and Michael Tyler, for giving me much-needed months to just write; Rusty Barnes and Austin Martin, for sharing their house; Reva, my dog, for the companionship and also her face; Jessica Case and Pegasus Books, for editing this book and making it happen; Zoe Sandler, my agent, for being a champion; Jack Welch, for putting up with my regular appearances in his living room; and Jill Tarter, for sharing her life.

INDEX

A

Active SETI, 5, 9. See also SETI

Agrawal, Avinash, 235

Alien-hunting telescope, 93–103

Allen, Paul, 72–76, 78–79, 187

Allen Telescope Array (ATA), 5–11,
   69–91, 93–103, 109–112, 219–221,
   228–230, 235–237, 248, 257

Alpha Centauri, 102, 247, 250

Alvarez, Luis, 134–136

Alvarez, Walter, 134–136

Ames Research Center, 6, 23, 81,
   108–109, 113, 123, 141–142,
   151–153, 170, 188, 211, 222–224,
   247, 252

Anders, Bill, 34, 81

Anderson, Chris, 229–231, 235

Antonio, Franklin, 98–99

Apollo, 190

Arecibo Observatory, 6, 8, 101,
   175–176, 188, 244

Arecibo Radio Telescope, 43, 177,
   203–205

Arrival, 223

Asteroids, 25, 135–136

Astrobiology, 99, 106, 193, 208,
   221–223, 254

Astrometry, 214

Astronomy, 4

Astrophysical Journal, 69

B

Back to the Future, 78, 121

Backus, Peter, 13, 166–167, 170–171,
   176

Bacteria, 210

Ballard, Robert, 210

Band waves, 6

Bandwidth, 6

Batalha, Natalie, 13–14, 217,
   222–225, 254–255

Baum, L. Frank, 60

Beasley, Tony, 99–100, 248

Berkner, Lloyd, 61

Bethe, Hans, 45–46

Bidwell, Debbie, 82–83

Big Bang, 24, 181

Billingham, John, 14, 105–106,
   112, 123, 128, 134, 139, 142–143,
   151–152, 177

BIMA array, 78
Biosignatures, 106
Black, David, 14, 124, 220
Black holes, 52–54, 78, 99–100, 126, 146, 149
"Black smokers," 210, 213, 218
Blueshift, 6
Bock, Douglas, 72
Borucki, William, 14, 81, 155, 215–218
Bowyer, Stuart, 14, 55–56, 66, 68, 105, 106, 123
Branson, Richard, 189
Breakthrough Foundation, 247
Breakthrough initiatives, 6, 247–249, 250
Breakthrough Listen project, 6, 8, 247–249
Breakthrough Message program, 6, 247
Breakthrough Starshot, 6, 247
Brin, David, 14, 238, 239, 240, 243
Broad, William, 202, 203
Brock, Thomas, 210
Brooks, Kate, 52
Brown dwarfs, 45, 67, 112, 123
Bryan, Richard, 15, 171–173, 178, 180
Burbidge, Margaret, 156–157
Burke-Spolaor, Sarah, 101
Bush, George W., 246

C
Calvin, Melvin, 63
Carbon detection, 211–212
Catmull, Ed, 140
Challenger, 179
Clark, Tom, 125–129
Clarke, Arthur, 252
Classical and Quantum Gravity Journal, 163
Clinton, Bill, 212
Cocconi, Giuseppe, 15, 58–60, 62

Cohen, Jim, 146
Columbus, Christopher, 141
Combat Zones That See project, 251
Confessions of an Alien Hunter, 202, 238
Contact, 2–3, 6, 76, 81, 91, 121, 144, 149–150, 204–205, 223, 229, 256
Conte, Silvio, 171
Cooper, Danese, 235
Corliss, Jack, 210
Cornell, Betty, 15, 29, 36–37, 42, 197–199
Cornell, Dick, 15, 28–31
Cornell, Ezra, 37
Cornell, Jill, 26–42. See also Tarter, Jill
Corson, Dale, 42–43
Cosmos, 134, 136, 245
Creativity, Inc., 140
Cullers, Kent, 15, 144–145
Curiosity, 211
Cuzzi, Jeff, 125–127
Cyclops project, 10, 65, 105–106, 123
Cyclops Report, 65–67, 105, 143, 176, 214

D
Dante, 210
Dark matter, 96, 149
Davies, Paul, 239
DeBoer, David, 15, 74, 200
Deep Space Network, 7, 143, 166, 170–171, 246
Defense Advanced Research Projects Agency (DARPA), 5, 251–253
Denver, John, 179, 180
DeVincenzi, Donald, 141–142
DeVorkin, David, 51
Diamond, Bill, 15, 220
Dish, The, 190
DNA, 168–169
Doppler effect, 7

Drake, Frank, 3, 7, 10, 15–16, 26, 56–63, 106, 125, 141, 151, 168–169, 184, 193

Drake equation, 7, 63–64, 152, 208–209

Dreher, John, 16, 182, 201–202

Druyan, Ann, 16, 149, 202, 243, 244

Duluk, Jay, 155–156

Duncan, John, 172

Dwarf galaxies, 51, 101

Dwarf star, 200, 219–221

E

Earth Speaks program, 246

ECHELON program, 77

Effelsberg Telescope, 137

Ehrmann, Max, 34

Einstein, Albert, 163

Ekers, Ron, 16, 190, 191

Electromagnetic radiation, 7, 52

Europa, 213

Europa Report, 213

Everitt, Francis, 163

Exobiology, 105–106, 175, 222–223

Exoplanets, 7, 88–89, 172, 207–226, 248, 254

EXPRES Project, 214, 215, 222

Extinction, 134–135

Extraterrestrial intelligence, 1–5, 23–26, 60–68. See also SETI

Extraterrestrial life, 88, 208, 235

Extremophiles, 7, 109, 207–226

F

FAST antenna, 249

Fast Fourier transform (FFT), 154

Fast radio burst (FRB), 100–102

Feed, 7, 94–97

Female scientists, 118–121, 224–226

Ferris, Timothy, 243

First SETI Protocol, 239

Fischer, Debra, 16, 214, 215, 222, 223

Fisk, Leonard, 165, 188

Follow-Up Detection Device (FUDD), 8, 192–194, 199–200

Fossils, 212

Foster, Jodie, 2, 6, 76, 81, 162, 205–206, 208

Fractalesque patterns, 212

Frail, Dale, 215

Frequency, 8, 26

Frequency of Earth-Sized Inner Planets (FRESIP), 217

Friis, Harald, 257

Friis, Inger, 257–258

G

Gamma rays, 53, 58–59

Garn, Jake, 17, 172–173

Garner, Donna, 108

Garner, Jack, 108

Gates, Bill, 75

Gender policies, 35–37, 43, 118–121, 185, 224–226

Gertz, John, 240–243

Goldin, Daniel, 17, 178, 188, 222

Gravity Probe B, 163

Green Bank observatory, 60–64

Green Bank Telescope, 1, 3, 8, 77, 125–129, 199–203, 248–249

Griffiths, Lynn, 155

Grinspoon, David, 238, 240

Grisham, John, 34

Gulkis, Sam, 17, 145–147, 177

H

Habitable zone, 8, 82, 217–219

Haines-Stiles, Geoffrey, 161

Handler, Philip, 134

Harp, Gerry, 17, 73–74, 121, 138, 154, 197, 203, 220

Hat Creek Radio Observatory, 8, 66, 69–74, 77–83, 89–91, 105–113, 121–123

Hawking, Stephen, 249
Heinlein, Robert, 48
Hewlett, Bill, 184, 186, 188, 195, 197
High-Resolution Microwave Survey
    (HRMS), 8, 9, 174–175, 181, 194.
    See also Microwave Observing
    Project
Horowitz, Paul, 249
Huntress, Wesley, 188, 222
Hydrostatic equilibrium, 70

I

Intelligent Life in the Universe, 55–56
Interstellar, 223
Interstellar communication, 59–60,
    106, 123, 152, 169, 240, 246
Interstellar messages, 168–169, 240,
    246–247
Interstellar travel, 25, 251–254

J

Jemison, Mae, 17, 251, 253–254
Jet Propulsion Laboratory (JPL), 8–9,
    139, 142–147, 166–167, 188
Jodrell Bank Observatory, 8, 145–147
Jordan, Jane, 177
Jorgensen, Susie, 84
Jupiter, 143, 213, 214, 243

K

Kapor, Mitchell, 187
Kelly, Kevin, 173
Kepler Space Telescope, 9, 81–82, 89,
    215–217, 219–224
Kepler spacecraft, 215
King, Ivan, 51–52, 117–118
Korbitz, Adam, 242–243
Kuhi, Len, 52–53

L

LA Times, 174–175
Labeled Release Experiment, 211

Lea, Susan, 118
Li Yao, 227–228
Liddle, David, 71
Life-detection experiments, 211–212
Light-years, 9, 24, 27, 61, 70, 90,
    100–101, 140, 220
Lilly, John, 63
Linscott, Ivan, 155–156
Lomberg, Jon, 243, 245
Lorimer, Duncan, 100–101
Lovell, Bernard, 8, 146–148
Lovell Telescope, 145–146

M

M dwarfs, 219–221
MacElroy, R. D., 210
Machtley, Ronald, 171
Manhattan Project, 46, 58, 253
Marcy, Geoff, 248–249
Mars, 59, 70, 150, 211–213, 230
Mars rovers, 211–212
Martian, The, 223
Masers, 2, 9, 144–147
Mass extinction, 134–135
McAuliffe, Christa, 179
McConaughey, Matthew, 205
McKay, David S., 212–213
McLaughlin, Maura, 100, 102
MeerKAT, 102–103
"Message from Earth," 246
Messages, interstellar, 168–169, 240,
    246–247
Messerschmitt, Dave, 155
Meteorites, 212–213
Methane environments, 106, 210, 213
METI (Messaging Extraterrestrial
    Intelligence), 9, 237, 242–243, 246.
    See also SETI
Microwave Observing Project
    (MOP), 9, 139, 162, 174–175. See
    also High-Resolution Microwave
    Survey

Mikulski, Barbara, 171, 173
Milner, Yuri, 6, 17, 247, 248
Mobile Research Facility (MRF), 9, 189–193, 199, 203
Monk, Greg, 129–130
Moore, Gordon, 9, 187, 195, 197
Moore's law, 9, 71
Mopra telescope, 10, 192–194
Morrison, Philip, 17, 58–63, 216–217
Mount Wilson Observatory, 156
Multichannel spectrum analyzer (MCSA), 143, 154
Murray, Bruce, 166
Murray, Elyse, 151
Myhrvold, Nathan, 71, 230

N
Nagel, Thomas, 169
Narrowband radio signals, 10, 60, 144
National Aeronautics and Space Administration (NASA), 10, 23, 34, 48, 65, 140, 173. See also Ames Research Center; SETI Institute
National Radio Astronomy Observatory (NRAO), 2, 10, 26, 61, 77, 100, 129–130, 193, 248
National Science Foundation, 43, 74, 108, 139, 153, 163, 189, 204
Nature, 60
Neller, Chris, 18, 202, 255
Neptune, 167, 243
New Horizons, 245
New York Times, 149, 169, 176–178, 181, 202

O
Obama, Barack, 246
Ögelman, Hakki, 41
Oliver, Bernard, 18, 62–65, 106, 151, 165, 167, 177–178, 181, 184–187, 197, 248

Oliver, Suki, 185–186
One Earth Message project, 245–246
One-Hundred-Year Starship, 5, 247, 250–253
Opportunity, 211
Optical telescope, 53
Order of the Dolphin, 10, 63–64, 106
Ozma project, 3, 10, 60–64

P
Packard, Dave, 184, 186, 188, 195, 197
Papadopoulos, Greg, 71
Parkes Radio Telescope, 10, 100–101, 171, 190–195, 249
Pasteur, Louis, 159
Patton, Jody, 187
Pearman, J. Peter, 62–63
Peterson, Allen, 155
Peterson's Left Leg, 154–155, 166–167
Phoenix project, 8–10, 72, 189, 197–204, 222, 249
Photometry, 215
Pierson, Tom, 18, 151–153, 183–184, 219
Pioneer 10, 167–169, 176, 194
Pluto, 70, 245
Poundstone, William, 60
Price, Mark, 191
Project Cyclops, 10, 65, 105–106, 123
Project Ozma, 3, 10, 60–64
Project Phoenix, 8–10, 72, 189, 197–204, 222, 249
Proxima Centauri, 219, 239, 250
Proxmire, William, 18, 138–139, 141–143
Pulsar, 215

Q
Quest for Contact, 161–162

R

Radio astronomers, 54, 99–102, 129, 145, 205

Radio astronomy, 1–2, 76–77, 99–100, 253

Radio frequency interference (RFI), 11, 145

Radio telescopes, 7, 11, 26, 43, 58–61, 66

Radio waves, 7, 11, 12, 53, 59–60, 76–77, 85–89

Randall, Karen, 80–81, 230

Raup, David, 135

Reason, 141

Receiver, 11

Red dwarfs, 221

Redshift, 11

Reykjalin, John, 143

Richards, Jon, 18, 74, 88–89

Roberge, Aki, 224

Ross, John, 189

S

Sagan, Carl, 2, 6, 18–19, 21, 55–56, 60, 63, 81, 134, 136, 142–143, 149–151, 202, 243–245, 250

Salpeter, Ed, 43–44, 46–48, 148

Salpeter, Miriam (Mika), 47–48

Saturn, 70, 143, 167, 213, 243

Schweizer, Linda, 52

Science, 143, 212

Scientist, The, 141

Seager, Sara, 222

Second SETI Protocol, 11, 239

Sepkoski, Jack, 135

SERENDIP projects, 11, 68, 72, 102, 105–108

SETI (search for extraterrestrial intelligence), 1–5, 23–26, 60–68. See also METI

SETI budget, 139–143, 151–152, 171–173, 178

SETI funding, 159, 162–164

SETI Institute, 1, 11, 34, 72–88, 97–112, 123–124, 139–144, 151–154

SETI Live, 11, 235–237

SETI Pioneers, 59, 61, 62

SETI Protocol, 7, 11, 239

SETI Quest, 12, 235

SETI Stars, 81

Sheaffer, Robert, 141

Shklovskii, Ivan, 55–56

Shostak, Seth, 19, 74, 202, 216, 220–221, 238, 245

Siemion, Andrew, 249

Signals, 181

Silk, Joe, 57, 112–113, 115, 118

Snowden, Edward, 77

Solar and Heliospheric Observatory (SOHO), 202

Space Sciences Laboratory, 134

Spectrometers, 52–53, 135, 145–146, 167

Spirit, 211

Sputnik, 251

Square Kilometre Array, 102–103

SRI International, 79–80, 94–96

Stanford Daily, 155

Starship program, 5, 247, 250–253

Steffes, Paul, 200

Struve, Otto, 62

Supernovas, 25, 54, 76, 99–100, 126, 140, 168, 215

T

Tams, Tim, 193, 194

Tarter, Bruce, 19, 40–43, 46–50, 54–57, 113–114

Tarter, Jill
background of, 1–11
cancer and, 196–198
career of, 66, 69–91, 93–103, 105–226
daughter of, 19, 50, 57–58, 113, 122, 125

early years of, 26–44
education of, 32–52, 55–56
life with Jack, 33–35, 44, 116–117, 123–131
lobbying efforts of, 162–164
marriage of, 42, 130–131
parents of, 15, 28–31, 36–37, 42, 197–199
retirement of, 73–74, 153–154, 208, 254–257
speeches by, 23–24, 229–235
Tarter, Shana, 19, 50, 57–58, 113–116, 122, 125, 130, 137, 157–158, 227–228
Tatel, David, 238
Tatel Telescope, 61
TED Talk, 229–235
Telescopes. See also specific telescopes
alien-hunting telescope, 93–103
optical telescope, 53
radio telescopes, 7, 11, 26, 43, 58–61, 66
Terzian, Yervant, 148
Thiomargarita bacteria, 210
Time, 62, 229
Titan, 213
Townes, Charles, 19, 249
Trimble, Virginia, 148
Trump, Donald, 241–242
Turnbull, Margaret, 19, 217, 219, 222–225, 255–256

U
Uranus, 243

V
Vaile, Bobbie, 193, 198
Vakoch, Douglas, 19, 169, 240, 243
Venera, 146–147
Venus, 136
Viking, 211

Von Braun, Wernher, 179, 180
Voyager, 143, 149, 243–245

W
Wavelengths, 12, 53, 59–60, 144
Webster, Larry, 170
Welch, Jack
background of, 19–20
career of, 66, 72–79, 87, 94–95, 105–113, 133–138
family of, 125, 137, 158–159, 229
life with Jill, 33–35, 44, 116–117, 123–131
marriage of, 130–131
Welch, Jeanette, 125, 137, 229
Welch, Leslie, 137, 229
Welch, Ruth, 158–159
Werthimer, Dan, 20, 72, 74, 79, 86, 102–103, 108, 248–249
WFIRST telescope, 222, 224
Wolszczan, Alexander, 215
Women scientists, 118–121, 224–226
Woodbury telescope, 199–203
Worden, Pete, 20, 23–24, 247
Wright, Shelley, 250

X
X-ray technology, 54, 59, 73

Z
Zaitsev, Alexander, 246
Zemeckis, Robert, 6, 204
Zuckerberg, Mark, 249
Zupan, Marty, 141